Making an Industrial Revolution

Making an Industrial Revolution

Skill, Knowledge, Community
and Innovation

Gillian Cookson

THE BOYDELL PRESS

© Gillian Cookson 2025

All Rights Reserved. Except as permitted under current legislation no part of this work may be photocopied, stored in a retrieval system, published, performed in public, adapted, broadcast, transmitted, recorded or reproduced in any form or by any means, without the prior permission of the copyright owner

The right of Gillian Cookson to be identified as the author of this work has been asserted in accordance with sections 77 and 78 of the Copyright, Designs and Patents Act 1988

First published 2025
The Boydell Press, Woodbridge

ISBN 978 1 83765 222 8 (hardcover)
ISBN 978 1 83765 141 2 (paperback)

The Boydell Press is an imprint of Boydell & Brewer Ltd
PO Box 9, Woodbridge, Suffolk IP12 3DF, UK
and of Boydell & Brewer Inc.
668 Mt Hope Avenue, Rochester, NY 14620-2731, USA
website: www.boydellandbrewer.com

A catalogue record for this book is available from the British Library

The publisher has no responsibility for the continued existence or accuracy of URLs for external or third-party internet websites referred to in this book, and does not guarantee that any content on such websites is, or will remain, accurate or appropriate

Contents

List of illustrations — vi
Acknowledgements — viii
Note on places — x
List of abbreviations — xi

Introduction — 1

1. What people knew — 8
2. Innovations in iron — 30
3. Enlightened engineering — 62
4. Gentlemen and players — 95
5. The reach of science — 113
6. The shape of networking — 142
7. Creating a scene — 171
8. Perspectives — 194

Appendix: Webs of connection — 210
Glossary of terms related to ironmaking — 226
Bibliography — 228
Index — 243

Illustrations

Map

1. Mainland Britain, indicating main locations referred to in the text (© F.T. Cookson) — xii

Plates

1. Extract from the Brearley Memorandum Book, 1773 (with permission of the Yorkshire Archaeological and Historical Society) — 16
2. Title page of Andrew Yarranton, *England's Improvement by Sea and Land* (1677) — 23
3. Darby's patent iron pot (with permission of the Ironbridge Gorge Museum Trust) — 42
4. Penry Williams, 'Crawshay's Cyfarthfa Ironworks', *c.* 1827 (with permission of Amgueddfa Cymru – Museum Wales) — 54
5. Dalton's symbols of the elements, 1806 (with permission of Wellcome Trust Corporate Archive, Creative Commons) — 68
6. Questions posed by Lewis to Cockshutt (with permission of Cardiff Libraries) — 79
7. Middleton Colliery transactions with Salt & Gothard, 1781 (with permission of West Yorkshire Archive Service, Leeds) — 96
8. Richard Hattersley's daybook, 1793 (with permission of West Yorkshire Archive Service, Bradford) — 97
9. Interior of an Alchemical Laboratory (with permission of the Wellcome Library, Creative Commons Attribution Licence) — 133
10. The Wortley district, from Joseph Dickinson's plan of 1750 (with permission from The Milton (Peterborough) Estates Co. and Sheffield City Council) — 149

11. Exchange of notes between Callisthenes Thomas and William Spencer, 1739 (with permission of Barnsley Archives and Local Studies)   160
12. Thomas Slaughter's Plan of Coalbrookdale, 1753 (with permission of the Ironbridge Gorge Museum Trust)   187
13. Plan of Crowley's Swalwell Works, *c.* 1728 (with permission of Tyne and Wear Archives)   188
14. Plan of Kirkstall Forge in 1816   191

The author and publisher are grateful to all the institutions and individuals for permission to reproduce the materials in which they hold copyright. Every effort has been made to trace the copyright holders; apologies are offered for any omission, and the publisher will be pleased to add any necessary acknowledgement in subsequent editions.

Acknowledgements

The theme of community has run through this work from its start, during the strange circumstances of pandemic. Being at times confined to home, and with access to libraries and archives severely restricted, proved to be a sharp lesson about how networks sustain us, and how much we take them for granted.

All this administered a push out of a comfort zone. The main focus of research, eighteenth-century ironmaking, was already daunting. But it was necessary, promising to explain apparent contradictions in the course of ironworking and the iron trade. And while technical issues are important, the crux is human connection. This suited a fresh perspective. The agenda, a long list of questions of my own, made few assumptions about where this would end. That approach fitted the zeitgeist, the uncertainties of the situation.

If there is to be a dedication, it is to the power of community. And so here, with my thanks, I acknowledge the relationships, the associations, which have made the difference.

I am especially indebted to two anonymous readers, whose comments were reassuring but also challenging, especially about how the work was arranged. That was invaluable.

I acknowledge with gratitude my fellowship and facilities in the School of History at the University of Leeds.

Fellow participants in the Arkwright Society conference organized from Cromford in November 2021, especially Joel Mokyr and John Styles, provoked reconsiderations. So too did joining the workshop 'Knowledge, Energy, and Industry in the Age of Revolution' at the University of Birmingham in November 2023 (Arts and Humanities Research Council Research Network 'Reframing the Age of Revolutions'). The contributions of Maxine Berg, Chris Evans and Pat Hudson were particularly pertinent.

I thank Barbara Hahn for guidance on unfamiliar (to me) aspects of the history of technology, for commenting on early drafts of chapters, and for being a good friend. John Suter continues to teach me a great deal about the technicalities of ironmaking, and generously shared many insights. To other friends too I am grateful for support, ideas, and encouragement: to Brian Barber, John Cantrell, Stephen Caunce, Colum Giles, David Jenkins, and Rachael Unsworth. For specific help, I thank David Hunter (West Yorkshire Archaeology Advisory

Service), Joanna Layton (on Ditherington Mill, Shrewsbury), Cormac Ó Gráda, Philip Riden, Göran Rydén, Edward Sargent, Geoffrey Tweedale, and Barry and Margaret Tylee.

Special thanks to archivists, librarians, and custodians who provided a service in difficult circumstances when visiting in person was not always possible: staff of the British Library, Boston Spa; Mark Isaacs and Gordon Anderson, Cardiff Libraries; Lucy Lead, V&A Archivist at the World of Wedgwood; Georgina Grant (and for added information on the Darby iron pot) and her successor at Ironbridge Gorge Museum Trust, Sarah Roberts; Tim Knebel and Robin Wiltshire, Sheffield City Archives; Katy Best, Barnsley Archives and Local Studies; Laura Gardiner and Dean Stockdale, Rotherham Archives; Neil Adams, Borthwick Institute; Alyson Pigott, Tyne & Wear Archives; Lucinda Yeadon of Commercial Estates Group, Kirkstall Forge; Vicky Grindrod, Fiona Marshall, and many colleagues at the West Yorkshire Archive Service; staff at Leeds University Library Special Collections, and Richard Hoyle and Sylvia Thomas of Yorkshire Archaeological and Historical Society for facilitating access and permissions for the Brearley Memorandum Book; Michelle Lewis and Kelly Powell, Cyfarthfa Castle Museum & Art Gallery; Kay Kays, Amgueddfa Cymru/Museum Wales; Iwan Ap Dafydd, Llyfrgell Genedlaethol Cymru/The National Library of Wales.

To anyone – and you are many – who has ever asked me a question I couldn't immediately answer: thank you – that was a great service.

Michael Middeke, editorial director at Boydell & Brewer, saw something in a much earlier version of this work. Long before it fell into place, he patiently kept faith, and has continued to encourage me. I am immensely grateful for that, and also offer thanks to Crispin Peet, Peter Sowden, Tracey Engel and Karen Francis for their invaluable help in bringing this to publication.

Thanks to my family – Joe and Gabriel, Frank (who drew the map) and Anna – for keeping me grounded and amused. The biggest thank you is for Neil, who has lived with this work in progress for a very long time. He rose to the impossible task with support, insights and suggestions, which encouraged, challenged, and refocused me.

Note on places

County names are given as they were before local government was reorganized in 1974.

In northern England, Lancashire, Cumberland, and Westmorland (on the western side) were divided by the Pennines from Yorkshire (to the east). Yorkshire was divided administratively into three ridings, of which the West Riding was by far the largest in area and population. It was also highly urban, with Sheffield, Leeds, and other industrial centres situated there.

Here, where 'south Yorkshire' is referred to, it means Sheffield, Rotherham, and Barnsley and their hinterland. It is not to be confused with 'South Yorkshire', a defined modern administrative unit. Similarly 'west Yorkshire' encompasses industrial areas around Leeds, Halifax, and Keighley, in a more general sense than the current 'West Yorkshire'.

'The Wortley district' is used as convenient shorthand for parishes in the upper reaches of the River Don, as explained in the caption for Plate 10.

The Black Country, now part of the West Midlands, centres on the town of Dudley, to the west of Birmingham. It has a long history of coal-mining and ironmaking, though its name seems to date only from the mid-nineteenth century.

The West of England included the historic counties of Cornwall, Devon, Somerset, and Wiltshire. In parts of those, and in neighbouring Gloucestershire, were significant textile industries. That was so, too, of East Anglia, comprising the counties of Norfolk, Suffolk, and Essex.

North-east England, the counties of Durham and Northumberland, is defined by the Tees valley to the south, with the Tweed marking the Scottish border to the north. Its centre is Newcastle upon Tyne, particularly associated with coal-mining and coal-trading.

Abbreviations

BAAS	British Association for the Advancement of Science (est. 1831)
BALS	Barnsley Archives and Local Studies Department
BDCE	Skempton et al., *Biographical Dictionary of Civil Engineers*, I.
BIA	Borthwick Institute for Archives, University of York
BL	British Library
BP	British Patent
Brotherton SC	Brotherton Library Special Collections, University of Leeds
DNB	*Oxford Dictionary of National Biography* (Oxford University Press, 2004, with online updates)
FRS	Fellow of the Royal Society
Lancs	Lancashire
Lit. & Phil.	Literary and Philosophical Society (but note, Leeds Phil. & Lit.)
OED	*Oxford English Dictionary*
RA	Rotherham Archives
RI	The Royal Institution [of Great Britain]
RS	The Royal Society ['The Royal Society of London for Improving Natural Knowledge']
RSA	Society of Arts; from 1908, Royal Society of Arts
RSE	The Royal Society of Edinburgh
SCA	Sheffield City Archives
TNA	The National Archives
T&WA	Tyne and Wear Archives
WYAS	West Yorkshire Archive Service
YAHS	Yorkshire Archaeological and Historical Society

Map 1. Mainland Britain, indicating main locations referred to in the text (© F.T. Cookson).

Introduction

Loose threads

The Age of Machinery concluded a long exploration of the origins of textile machine-making. It was a study of innovating, in workshops and on the factory floor, describing how a new trade was created which paved the way for modern mechanical engineering. The focus was on mechanization's epicentre – the textile towns of northern England, from the late eighteenth century to the mid-nineteenth. When published in 2018, it filled a gap in the story of British industrialization.[1]

That could have been the end, except that significant loose ends had emerged. There was a bigger picture, more still to be said about industrialization's broader context – other themes, other places, and extended connections which came to light in the course of research. Here is the result, and it is a different kind of book. The subject matter extended into earlier periods and less familiar industrial pursuits, so an exploratory and inquisitive approach seems appropriate. This was an opportunity for open-mindedness, and certain findings indeed proved sufficiently startling, or at least unexpected, to prompt some rethinking.

Relationships are at the heart of this new work. It takes further the emphasis on connection in the machine-making research, and chases down hard evidence about the nature of those associations. It shows how personal links, tapping into channels of experience and learning, fed into industrial innovation. While knowledge, embodied in communities as well as in individuals, is barely quantifiable, it can be recognized and described.[2] Without doubt, innovation was widespread and it came in many forms. The places where making and trading occurred are especially rich territory for reaching new understandings of industrial change.

While institutions had a part in spreading knowledge, that should not be overstated. Victorians who wrote on these events were in many cases mesmerized

---

1 Cookson, *Age of Machinery*.
2 Here, 'knowledge' is not understood as synonymous with 'science'. Secord, 'Knowledge in Transit', discusses how the two relate.

by formal institutions, missing the point about how things had been done a century earlier, before those bodies existed. To best understand the impetus behind technological and other developments, the eighteenth century has to be appreciated in its own time, and through an awareness of industrial relationships within communities. The concept of 'Scenius', described in chapter 7, underscores the boundless creative potential of human beings as they gathered around the birth of something new.

Taking a fresh and questioning approach, giving fair dues to workshop experience and manual skills – from this, a case builds that these practices far outweighed science in shaping eighteenth-century industrial technologies. This presents a challenge to Mokyr's 'industrial enlightenment'.[3] The modern reader needs to appreciate what 'science', as practised and understood by contemporaries, actually meant at the time. What exactly was it? And then, through what channels did it influence industrial innovation, if indeed it did? Contemporary meanings of many other terms and concepts are equally important. So too is the extent to which social class determined what was accessible and possible for individuals.

The reign of William III brought financial innovations, notably the Bank of England's establishment by royal charter, as the state looked to develop commerce and promote overseas trade.[4] Regarding the technological revolution in progress, though, government took little notice, other than in matters of taxation. New products, notably armaments, would later attract its attention. But little suggests that the British state was interested in encouraging innovation.

There is one case where state funds were influential in seed-funding innovation. The results were remarkable, but so were the circumstances, for it was an outcome of the Darien disaster which wrecked the Scottish economy in around 1700. This is described in chapter 7. For political reasons, Scotland received modest funding to support certain of its industries after its absorption into the United Kingdom in 1707. Not much happened for twenty years, but then university-based scientists spotted the potential for industry-based research. Through a mix of historical accident and religious circumstance, the small nation went on to carve a unique and influential place in British industrialization. The Scottish phenomenon deserves to be better recognized, and not only as a fascinating counterpoint to events elsewhere.

So some of the stories and characters under discussion here are rather less familiar than others. The appendix is an assemblage of individuals mentioned in the chapters, grouped by provable connection. Some were not very notable, nor necessarily successful. But the exercise demonstrates how individuals, well known or not, linked into circles of achievement, great or small, within or beyond

---

3   Mokyr, *Enlightened Economy.*
4   And particularly, it is argued, the slave trade: Berg and Hudson, *Slavery, Capitalism and the Industrial Revolution.*

regions, some interested in science or faith, some in profiting or profiteering, in creativity, in family. The appendix is simple, cutting through complexities to indicate how things, people, thoughts, travelled around and grew.

Iron has a big part in all this. Understanding the progress of ironmaking and iron-trading over the long eighteenth century is a priority in an endeavour of this nature. The switch towards iron-framed machines started in the 1790s, as iron replaced wood in machine tools and textile machinery, and for other applications in mechanical and civil engineering. Before that, over the preceding century, the changes in iron manufacturing and iron products influenced the course of many other industrial endeavours. Of special interest are the problematic decades before iron-puddling became viable. The story of iron weaves in and around any account of British industrialization.

Iron is a connective element: physically, into other industries, as product or material; with science, whose development it informed; and in setting templates for how early textile-engineering and other industries organized their own operations and working relationships. Its advance had effects which were protracted and transformational on many fronts. This connectivity produces evidence of the social glue that brought together people and projects. It enables more nuanced views of how that world worked. Proximity, we will see, cannot in itself prove influence. Nor can it be assumed that information, just because it existed somewhere, was known everywhere, or even elsewhere. Older assumptions and recycled sources need to be challenged.[5]

This is why a weight of evidence counts. The family and trade connections in ironworking, engineering, and merchanting were key. Nothing was static, and those relationships and creative hubs, and a lot of quite ordinary people, were instrumental to finding industrial solutions.

Aims and approaches

At the centre is a question about how knowledge and connection fed innovation. Where did useful knowledge originate, and how exactly was it disseminated?

Of the structures nurturing innovation, some were abstract, others physical. At work were personal and industrial links which were not confined by place, and this story embraces British regions and localities which have not featured large in accounts of industrialization. To an extent, it also covers events of relevance in Britain's international trade, including connections with North American colonies, Sweden, and other parts of continental Europe.

---

5   Musson and Robinson's *Science and Technology* remains influential, and fifty years on is still widely quoted. A compendium of articles, it offers useful primary material but questionable conjectures. See D.C. Coleman in *Econ. Hist. Review* (1970).

Reappraising the industry–science dynamic delivers fresh and persuasive evidence to a long-standing debate. Science developed in the eighteenth century. Knowledge developed. Experimentation developed. But how – where, when, in what shape – did those pursuits support the industrial technologies then multiplying? How relevant was science, as it stood in the eighteenth century, to industry, how exactly to new technology? Where the worlds of science and industrial innovation met and shared information, if that is indeed what occurred, has to be identified. As for which of those two spheres made the net gain from information exchanges, nothing is to be assumed.

So this must be robust, substantiated as far as contemporary sources allow. Some claims of knowledge transfer do not survive a brush with chronology. Or they do not bear scrutiny, as with the relationships of Papin and Savery to Newcomen; or they lack adequate detail, as with Desaguliers' connecting into provincial industry.[6] Sources are not always consistent with how historians have reported them.[7] An encounter, a connection, has to be verifiable.

Contexts made innovation, and innovation made contexts. Technology alone does not explain companies' pursuit of particular strategies. Sometimes there was sound reason – perhaps because of location, or because its owners had different plans for the long term – for a business to stick with an older technology. Technology should, though, be better understood by historians. Forty years ago, Rosenberg, seeing that economists could learn from knowing more about technical change, coaxed technology out of its black box.[8] Technology can similarly enlighten and enliven questions of history.

It is not too daunting: eighteenth-century science and technology are relatively uncomplicated. More problematic can be the challenge of business sources, which are more likely to be inconveniently scattered than housed in a company archive. Reconstructing these histories means extensive searching, for correspondence with landlords and associates, probate filings, notices in newspapers, family records, and whatever other incidental material is to be found. Supplemented with reliable secondary material, biographical sources, and background on regional industries and commerce, and set into a framework of eighteenth-century iron production and contemporary science, there are answers. The quest is far from hopeless.

Featuring here are three pioneering Yorkshire ironworking businesses which, in common with other small family start-ups, have almost no surviving records of their own. Yet the volume of material eventually unearthed was too much to fully accommodate here. Instead, complementary articles have already been published in regional journals about Hunslet Foundry in Leeds, and the

---

6 See pp. 135, 138.
7 For instance, the misunderstanding of the Gott notebook of mill practice (Brotherton SC, MS 193/117A), and John Marshall's records (Brotherton SC, MS 200). See p. 69–70.
8 Rosenberg, *Inside the Black Box*, preface, and *passim*.

Cockshutt family of Wortley Forge, near Barnsley. The Walker ironmasters of Grenoside and Masbrough are on a similar path.[9]

The past as a foreign country

Is it a surprise to discover that 'the Enlightenment' first appeared in English usage, capitalized and with the definite article, as late as 1893? So whatever eighteenth-century people imagined they were experiencing, it was not precisely 'the Enlightenment'. Another term, 'The Age of Enlightenment' was in circulation long before that, echoing the French *le Siècle des Lumières*. In France, *lumière* has come to mean a person of intelligence or knowledge.[10] Here is a reminder to 'mind our language'. That means more than recognizing anachronistic words as such. Historical language is a way into seeing what its users had in view.[11] So in the time of enlightenment, the *lumières*, the idea of illumination, offered a sense of fluidity and dynamism. A century later, 'The Enlightenment' as Victorians saw it leaves an altogether more stolid, institutional impression.

These matters – language and anachronistic thinking, and particularly the question of what was understood as 'science' – are further discussed in the opening section of chapter 3.

Industrialization, a protracted process, calls for an appreciation of cultures across generations. Developments in the eighteenth-century iron trade were rooted in the period of Civil War and Commonwealth, so this extends over a long period of accelerating change and great upheaval. And we have to ask, too, how much insight did early Victorians have into a sometimes distant world of industrial innovation during the previous century? That is open to conjecture. Even those experiencing workshop innovation during their own lives can have known little of how their great-great-grandparents' generation had confronted shortfalls in knowledge in earlier times.

This question, 'what did people know?', is at the core of appreciating industrialization. Addressing it means distinguishing between 'the problem of understanding for historians' – so, how we discover past events, and explain them to a modern readership – and a separate concern, about the contexts within which scientific and technological knowledge grew. That is, how do we find evidence about 'the problem of understanding that confronted the scientists historians [write] about'?[12]

---

9   Cookson, 'Hunslet Foundry' and 'Wortley Forge' are available on Open Access. A new DNB entry on the Cockshutt family was published in 2024.
10   The OED tracks changing definitions, and how contemporaries understood words. See also *Le Robert Dico en Ligne*: https://dictionnaire.lerobert.com/definition/lumiere.
11   Quoting Susan Reynolds, talking about medieval words in 2008 for the 'Making History' project: https://archives.history.ac.uk/makinghistory/resources/interviews/Reynolds_Susan.html.
12   Shapin, 'Paradigms Gone Wild'.

Because, relative to what would immediately follow, eighteenth-century strivers for illumination were much disadvantaged, grasping for tools to understand and explain processes and technologies. Such information, about that search for insights and ways to systematize their thinking, is of great interest. How were scientific and technological enquiries articulated without, for instance, a periodic table of elements? With no manual for techniques such as iron-puddling, which centred on skill, not hardware? What depth of understanding existed about, and what were the possibilities to articulate, activity witnessed on the workshop floor or in the laboratory? What, in short, did they think they were doing?

Their quest – broadly, innovation in science and technology – was creativity in action. It amounted to a significant work in progress on many fronts. It was not, however, predicated on accumulating knowledge in order to find a great truth. Unifying theories seem not to have figured much or at all. Theories and practices circulated, methods and concepts were absorbed and understood by those working in various settings, whether laboratory, workshop, or other. Change was endemic but, it seems, the general perception was of something piecemeal.[13]

A lot rests on proving connections within this upsurge of activity. By focusing on the scenes of engagement, working relationships can be more clearly defined, along with the physical realities of work. What kinds of resources were then available, and what were the technical constraints? How were things done, and how were those endeavours understood by contemporaries? Above all, the search is for wider connections, and the channels through which knowledge (broadly defined) passed.[14]

Content and chapters

The first three chapters relate to contexts of industrialization. Chapter 1, 'What people knew', suggests a baseline: how information and experiences had been noted and shared in quieter times, before the pace of innovation intensified over the century's middle decades. Possibilities for knowledge exchange expanded as regional and overseas connections grew, underlining the importance of merchants, industrial and trading links, and transport projects in disseminating information.

Chapter 2, 'Innovations in iron', follows the iron trade over a century of war and technical challenges: the relationship with Sweden, including the open exchange of technical information; innovations such as the gradual introduction of coke in iron-smelting, and the more decisive change delivered by the puddling

---

13 Ideas drawn from Shapin's discussion of Kuhn: Shapin, 'Paradigms Gone Wild'.
14 See Secord, 'Knowledge in Transit'.

process. The chapter addresses apparent contradictions in the view of British iron production stalling technologically after 1750 and before the advent of puddling, and opposing suggestions that the industry was in fact booming.

Chapter 3, 'Enlightened engineering', considers the association between science and innovative industries, defining something of what 'science' meant, and what constituted the scientific establishment.

The second section of the book concerns stepping up to challenges and develops arguments about the importance of skills and networks. Chapter 4, 'Gentlemen and players', addresses two aspects of industrialization which tend to be disregarded or misunderstood. Social class was the greatest determinant of access to certain kinds of education, training, work and finance, and consequently entry into certain networks and the possession of knowledge and skills. And then the chapter confronts the habitual misrepresentation of millwrights. 'Millwright' comprises a range of grades, skilled or professional, and industrial specialisms, some of which were highly influential in industrial development, though others much less so.

Chapter 5, 'The reach of science', considers what industry might have derived from universities, provincial and London-based learned societies, and public science events. It identifies a wide range of such groups, challenging historians' tendency to lump them together as 'Lit. & Phil.' societies. It also details the limitations of the vaunted cosmopolitan John Theophilus Desaguliers, placing him firmly in context.

Physical connection is the subject of chapter 6, 'The shape of networking'. This explores the industrial framework of English and Welsh regions, and the significance for innovation of migrants' skills and knowledge. Migration itself relied on the passage of information about opportunities elsewhere. What propelled skilled individuals to move? The chapter also looks at changing organizational models in ironworking.

Chapter 7, 'Creating a scene', sets up a picture at odds with Carlyle's 'great men' and Britain's mythical national story. It suggests that, instead, the innovation of this period should be viewed as creativity, and due credit assigned to collaborative working across many 'scenes'. It explores the real-world links between ninety or so individuals of acknowledged significance, and also considers why Scotland did so well, and was disproportionately present and influential.

The concluding chapter, 'Perspectives', has something to say about Adam Smith and economists. It then draws concluding points about innovation in the period of industrialization.

Finally, in the appendix, are listed those approximately ninety individuals of influence and interest, with a focus on their connections, and with references to the *Dictionary of National Biography* and *Biographical Dictionary of Civil Engineers*.

1

What people knew

What did people know – and how did they know it?

Industrial development demands new kinds of knowledge, while simultaneously generating it. Through doing and experiencing, knowledge emerges and is shared. It is a product of manufacturing and of commerce, created as systems and constructions, processes and products are developed. It can be a skill or a technique, manual or theoretical. It might be a combination of some of these, or of other things. The question encompasses great sweeps of industrial and intellectual endeavour. While it is apparently a simple one, the answers are far-reaching and promise to be complicated.

British industrialization is often identified with northern textile towns, where its effects after 1800 were spectacular. But it was in the eighteenth century that all this started, with a gradual systematization of discoveries, some of those of even earlier date. It was a slow but accelerating process. For each of the key industries and activities – the concern here is especially with iron and engineering and the theoretical work which informed those trades – it transpires that there is much uncharted ground to explore.

Industrialization was about rather more than locations and hardware. It ran on connection, facilitation, communication. Well-established trade links between regions played a part. And while industrial concentration increasingly focused on Lancashire, Yorkshire's West Riding, and the English Midlands, prolific and significant activity continued elsewhere. The variations in regional economies and industries are clear, and relative positions shifted over the century. All of this has to be added to the mix of influence, because the places later seen as driving forces were not the only ones where unique and interesting things had happened. At the very least, those others provide valuable context.

That textiles held a special role in industrialization is easily seen. Relentless demand from textile manufacturers was a big factor in driving mechanical innovation and machine production, and factory-building and all that went with that. Those forces accelerated through the second half of the eighteenth century and into the nineteenth. Yet certain aspects of textiles' contribution to the process were more subtle, and are less remarked on. Many advances were made before the notorious 'wave of gadgets'.[1] Textile interests constructed a commercial and technical framework which enabled Britain to industrialize as it did; much of that foundation was laid down before 1750; and it was a transregional phenomenon, significantly enabled through accrued wealth from the textile trade.

Groups and individuals who took the initiative – from early in the 1700s – to support trade, by creating a commercial context and physical infrastructure, could have had little sense of the longer-term implications of what they were engaged upon. The most obvious of such projects, those delivering the more dramatic effects on local landscapes, were transport improvements: canals and river navigations, turnpike roads, waggon-ways, ports, and harbours. Other kinds of scheme facilitated marketing and commerce: cloth halls, banking and insurance facilities, property registers to protect lenders. There was particular emphasis on exports of textile and iron products, though that was also a two-way traffic, bringing in raw materials and consumer goods and, most significantly, delivering coal at much reduced cost to expanding towns and industries. The courage to invest was bolstered by those growing trade possibilities, especially with the North American colonies. For the projectors, their own, fairly immediate, industrial and commercial needs were paramount, and centred in their own region or sub-region. However modest the aspirations, once achieved they formed stepping stones to more dramatic transformation.[2]

Channels of experience and learning

Roads and waterways were huge undertakings, demanding and expensive to build. Their very existence is confirmation, if confirmation were needed, of the size and importance of trade between regions, and internationally, even before 1700. A linear venture cannot be purely local. It must be negotiated along its course, planned, designed, land acquired, finance raised. There has to be some common purpose, understanding, and investment from the benefiting communities – and that last word can mean something beyond one locality. Communality came through working relationships, a necessary sharing of technicalities, business linkages, strategies, sketches, ideas, as well as the humdrum chatter of everyday life. 'Community' might embrace, to take one important example, participants within a specialist trade who were based in different places, perhaps even in different regions or nations.

---

1   To use Sellar and Yeatman's famous definition of the industrial revolution in their satire *1066 and All That*, which places full emphasis on the tangible.
2   For a synopsis of the Manchester region's transport developments, see Hahn, *Technology in the Industrial Revolution*, 91–2.

Exactly how these (and other) specialist networks – ones at the cutting edge of industrialization – functioned and interacted is a question for the following chapters. But where did all this start? How did people, and regions, make meaningful connection before the full-blown industrial take-off? An already wide view of the world, and of trade, and strong trans-regional connections, can be shown to have existed. This is important, for it demonstrates that knowledge had many opportunities to flow through channels which crossed regional and industrial boundaries. It was natural that merchants, and others who managed and organized trading, had interests that centred on their home territory. But their horizons necessarily extended far beyond their own place, beyond their region, and, in some instances, overseas. The working of supply chains, a grasp of which products were currently in demand, and where, and at what price – understanding and satisfying all of that was the route to success as a merchant.

Some of the proposed, or contemplated, infrastructure was minimal and localized. But other plans were large, ambitious, and very visible, calculated to ease the flow of regional and international connection. Where such schemes were projected and completed, that in itself suggests that substantial trade links already existed, sufficient to even think of funding such a vast undertaking. Maw, on Manchester, where the water transport age began in 1721, argues that such major undertakings were a response to growing commercial demand, and not in themselves stimulators of economic development.[3] This appears to be what occurred in Yorkshire's textile districts also. Naturally, the new facilities did support further expansion, and where traders linked, information would pass.

The earliest examples of substantial trade links come, unsurprisingly, from textiles. It is clear that long-distance trading was well established in early-modern times. A large-scale woollen-cloth production and marketing chain, in full flow by the 1580s, is described by Newton and backed by an impressive range of archival sources. Wakefield merchants were at its pivot, and London markets its ultimate destination.

That trading route was sufficiently settled and robust that, as early as 1621, there came the first suggestion for a river navigation to improve the movement of goods between Wakefield and the Humber. The idea returned from time to time, and finally took shape from the 1690s in a wider, sub-regional scheme, the Aire and Calder Navigation. That stage, Wakefield to Hull – a particular problem, it would seem – was one link in an important manufacturing and trading system starting with suppliers of wool and other goods from Cumbria, via the markets of Kendal, and from the Yorkshire Dales. The raw material passed by road to Wakefield, where merchants allocated it to agents and out-workers, there and in

---

3 Maw, *Transport and the Industrial City*, 1, 8.

neighbouring towns. Finishing processes, the cropping and dyeing which added such value to finished cloth, were carried out under the watchful eye of the merchants, often in Wakefield itself. Then, mainly, it was dispatched to Hull and onwards by sea to a specific cloth hall in London. A smaller trade in Wakefield cloth fed shopkeepers and merchants in towns on the Welsh borders – including Shrewsbury, where further dyeing and finishing was sometimes carried out – and in the West Country and Midlands and other places en route.[4]

The geographical spread of these networks, their early date, and the texture of relationships revealed by Newton's sources are striking. The Wakefield-based example illustrates the possibilities – indeed, the necessities – for sharing know-how along such trading channels. Personal connections sustained through generations, especially between merchant families, were often cemented by marriage, by loans, or in joint ventures such as those infrastructural projects. Relationships between regions, or sub-regions, could be very tangible. And thus, knowledge flowed.

Merchant-organizers in that Wakefield-centred system were highly specialized: they kept the manufacturing processes tight, near enough to the town to be monitored, while the most demanding finishing stages were supervised on their own premises or close at hand. They or their representatives had long-standing commercial, technical, and personal interactions throughout the chain, from north-western rural wool suppliers to the London merchants who fed back on demand in the markets, including exports. They were also highly attuned to their own town's vibrant economy. High-end knowledge – here, perhaps to be defined as experience and information which oiled the wheels of this trade and helped it advance – moved in multiple directions.

Change, then and later, was not confined to the textile industry, nor to the regions often most associated with industrialization. But the range of practice, the ways in which things were done in different places, needs to be recognized. Even within regions there was marked specialism. Halifax, for instance, was from 1600 a rising force in textiles, and its cloth-exporting expertise later transformed the shape of merchanting in Leeds.[5] As for the regions themselves, there were marked differences in how their economies worked, some of that attributable to the features of particular industries and local commerce. The classic success stories centring on parts of Lancashire, Yorkshire, and the Midlands ought not to cloud the prolific and significant activity happening elsewhere.

The sectors fundamental to industrialization, key to creating and developing innovations in their own and other trades, were ironmaking, textiles, engineering, and fuel extraction. The spread of those industries in fact embraces much of Britain, and very many of its towns: large areas of northern

---

4 Newton, 'Wakefield, Its Woollen-Cloth Trade and Merchant Networks'; Willan, 'Yorkshire River Navigation'; Unwin, 'Leeds Becomes a Transport Centre'.
5 Newton, 'Wakefield, Its Woollen-Cloth Trade and Merchant Networks', 136–7.

England, including Derbyshire and Cumberland (identified with iron, coal, and textiles); the Midlands, reaching to the Welsh border, and into Shropshire and Staffordshire (iron, steam engines); areas of Wales (iron, coal); Tyneside and Wearside (iron, coal, mining engineering); the West Country (textiles); and Cornwall (mining engineering). Scotland's prominent role in advancing all those industrial sectors will be further discussed. Multiple, highly significant, technological advances were made in London in a wide range of trades; in the stocking and lace-making towns of the east Midlands; and in engineering in the southern English dockyards.

But how exactly did regions connect with each other? Trade links were evident, necessary, physical associations. How, though, is knowledge to be tracked? What exactly passed, and how was it transmitted? And then how was it received, and what was done with that information? Was an active and purposeful quest for knowledge in progress, with specific problems targeted? Certain collaborative, cross-regional developments have been mentioned: just how common was that phenomenon? Knowledge transfer must often have been quite random and haphazard, as evidenced by the eighteenth-century commonplace books kept by John Brearley and others.[6]

Not all regional industries were able or inclined to embrace change. The East Anglian worsted trade, for instance, its long supremacy already weakening by the 1740s, was later left with few options as the market share for fine products collapsed. As Norwich merchants' own technical knowledge became obsolete, that of mechanized spinners – with whom they were in close contact – proved of little use, for they were quite unable to emulate what the Yorkshire industry was doing.[7] East Anglia was deficient in the natural assets increasingly demanded – coal, water power. Nor did it have the necessary proficiency to keep abreast of developments, new kinds of knowledge embodied in a trained and experienced workforce. Eventually there was no prospect of attracting such skilled labour, because there was no mechanical scene there, no context of automation.

The resources in universal demand – whether transportable or needing to operate in situ, such as water power or charcoal-burning furnaces or winning certain minerals – combined with the need for workers with relevant expertise, would determine how trade was organized.[8] Finding ways to work around such situations – introducing steam in place of water power, or ironworking techniques that could use mineral in place of vegetable fuels – in itself proved

---

6 See pp. 15–16.
7 James, *History of the Worsted Manufacture*, 223, 231; Berg, *Age of Manufactures*, 104–10; Wilson, 'Textile Industry'; 'Supremacy of the Yorkshire Cloth Industry'; Cookson, 'City in Search of Yarn'. For the medieval emergence of textile out-working, and an East Anglian context, see N. Amor, 'The Origins of the Putting-out or Domestic System of Industrial Production in England', in Nigro (ed.), *Knowledge Economy*, 263–85.
8 Berg, *Age of Manufactures*, 100–3.

a great stimulus to industrial innovation. But it could not offer a universal solution, for certain regions and localities were ill-fitted to respond.

As noted, much of Britain played some part at the vanguard of innovation, nurturing key changes in a range of industries. Because significant industries spread across more than one region, specialist trade knowledge would find its way around. Ironworkers exchanged practices, techniques, and other developments as they moved between jobs and localities, to wider benefit. Thus workers themselves were channels, and so were contractors and advisers in various important industries.

If everywhere, or almost everywhere, is potentially of great interest, here is encouragement to spread the investigative net much wider. That approach also reveals fascinating connections, personal and institutional. Questions along the lines of 'Why, considering its early advantages, didn't Shrewsbury (or Trowbridge, or Norwich) retain a greater role in textiles?' are not the most enlightening. It is more interesting to explore what was actually happening in such places, and to connect those local experiences into a wider picture. Because every place that 'failed' had an afterlife. Localities facing changed regional circumstances and economic pressures had to work out contingencies or substitutions.[9] These may not have proved successful, but whether they did or not, they hold significance in the overall story. Cases of localized economic decline could in fact be only relative within the context of industrial boom. New opportunities would materialize, and the displaced merchant or small manufacturer would ask themselves, 'What is now possible?' What was appropriate and practical? This is more nuanced than seeing certain regions triumph at the expense of others.

Change was widespread, and failure attended it daily. Even in the thriving regions, extruded manufacturing structures that had been less well managed than the Wakefield example, run by strings of smaller merchants and clothiers, agents and sub-agents, 'putters out' of work with little expertise within them, eventually proved unsustainable. Their great contribution to industrialization, it could be argued, was in collapsing. In doing so they delivered an incontestable impetus for change. Because the quality of management and workforce was poor – short on aptitude, training, motivation, and direction – it is implausible that much useful know-how had been generated in those far-flung extremities. The lesson here is not to assume that everything that was part of a markedly 'innovating' industry in a dynamic region was worth sharing.[10]

The mission, then, is to identify some of the more substantial flows and exchanges. Those could be significant transfers of knowledge which inspired innovation, or lesser events which were nevertheless effective and valued

9 Berg, *Age of Manufactures*, 110–12.
10 Cookson, *Age of Machinery*, 25–8.

in industrial communities. John Smail, historian of Yorkshire's woollen manufacture and its merchants, saw in the eighteenth century 'a process of cultural transformation that involved the interplay of economic and social change'.[11] This, above all, is what mattered, and what should matter to historians. As always, it comes down to people, because it was in the 'there and then' that decisions taken, relationships forged, laid courses which guided industrial transformation. This has been shown in the development of northern textiles and early textile engineers.[12] It remains to be demonstrated in a wider context, of industry, learning, and locality.

Merchant knowledge

To merchants and other intermediaries in trade, making connection was central to their mission. The role was not purely commercial: many merchants had involvement in manufacturing, and all were required to understand their product. They might deal in luxury goods or, more commonly, in everyday needs. Merchanting was integral to the smooth running of important branches of mass manufacturing and its profits – and people – sponsored infrastructure and developed urban centres of trade. Every branch of merchant activity rested on an array of contacts, and an appreciation of their value.

Eighteenth-century merchants were often to be found at the apex of a production system, overseeing – though not necessarily in full control of – a pyramid which rested on a base of large numbers of people working at a very small scale. Not all merchants were of equal consequence or competence; nor were the agents and other middlemen involved in transacting these businesses. But the nature of this work brought them into contact with quantities of information. This, while varying in its value, was fundamental to trade. To many trades, in fact, and merchant knowledge was central to industrial as well as trading activity.

Information passed by word of mouth and letter along the same channels as the merchandise. Daniel Defoe noted, on crossing the Pennines in the 1720s, that the Postmaster General had established 'a cross-post through all the Western Part of England into [Yorkshire] to maintain the Correspondence of Merchants and Men of Business, of which all this Side of the Island is full'. That service extended from Plymouth to Liverpool, and then via Manchester to Hull.[13] A significant long-distance trading route was being recognized and confirmed.

11   Smail, 'Manufacturer or Artisan?', 808.
12   Cookson, *Age of Machinery*, ch. 6.
13   Defoe, *Tour thro' the Whole Island*, Letter 8, Part 3.

Some of what passed between traders or tradesmen would be ephemera, quickly forgotten or distorted in the telling and retelling. Undoubtedly, much of the content was not of the highest quality or importance – the term 'knowledge' would flatter it. But news was power, and gaining significant intelligence about developments – whether local or further afield, about company or bank failures, who was in or out of business, prices of commodities, gluts and shortages, styles and fashions, availability of workers – all of this conferred advantage.

Judging by the commonplace books kept daily by the Wakefield cloth frizzer John Brearley, knowledge transfer had grown to a deluge. Brearley completed at least thirteen books of notes and sketches between the 1750s and 1770s. The contents related to his own and his clients' trades, and to the workings of his mill and waterwheel, including technical sketches. He also ranged far and wide with his own and others' scraps of news, recipes, and musings. Public mills like Brearley's, and other industrial meeting places, had become informal knowledge exchanges.[14]

Brearley, and similar examples of industrial information being noted in Yorkshire, date from the mid-century or later.[15] This is interesting, as the region's progression towards becoming a force in woollen-cloth making had, according to Smail, entered a second phase around 1750. The first involved a growth in the *scale* of production and trading; then, from the mid-century, in entering the fashion market, a change in *scope*, which introduced fresh approaches to marketing. That in its turn generated a new focus upon production systems, and an energy for improvement which fed into process innovation. Smail thus makes a chain of connection from the export trade back into manufacturing districts, which in consequence became the scene of radical change in technologies, meeting urgent new demands. And what had changed over the course of that century? Yorkshire merchants and larger manufacturers had achieved greater control over cloth production – including more direct access to markets, gaining privileged insight into the conditions there – better knowledge, in other words.[16]

And this shifting industrial scene came to encompass much more than the textile trade. By the 1790s, while textiles had come to dominate British exports, iron products were in second place, and remained so.[17] The iron trade, with its own networks of producers, financiers, and middlemen, experienced many

---

14 Smail (ed.), *Woollen Manufacturing in Yorkshire*, is a part transcript of WYAS Leeds, WYL463 (2 vols, 1758–62). A further Brearley commonplace book (1772–3) was acquired in 2021 by YAHS: see Brotherton SC, YAS MS2022. Brearley and other noters of technical information are further discussed in chapter 3.
15 Cookson, *Age of Machinery*, 162–4.
16 Smail, 'Causes of Innovation', esp. 2, 4–7; also Maw, *Transport and the Industrial City*, 235.
17 Jones, *Merchants to Multinationals*, 19.

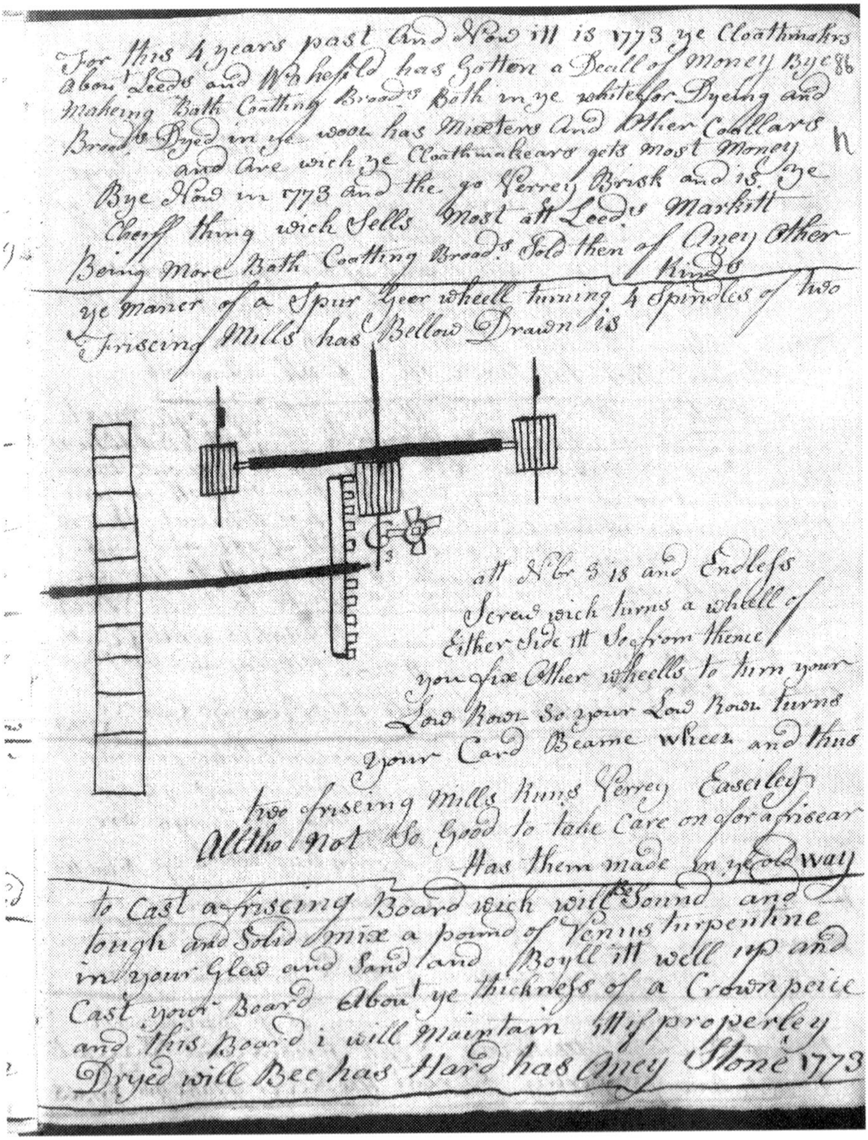

For this 4 years past And Now itt is 1773 ye Cloathmakers
About Leeds and Wakefield has Gotten a Deall of Money Bye 86
makeing Bath Coatting Broads Both in ye white for Dyeing and
Broads Dyed in ye wool has Mixters And Other Collars
and Are wich ye Cloathmakears gets most Money
Bye Now in 1773 and the go Verrey Brisk and is ye
Cheiff thing wich Sells most att Leeds Markitt
Being More Bath Coatting Broads Sold then of Aney Other Kinds

ye maner of a Spur Geer wheell turning 4 Spindles of two
Frising Mills has Bellow Drawn is

att Nbr 3 is and Endless
Screw wich turns a wheell of
Either Side itt So from thence
you fix Other wheells to turn your
Low Rowr So your Low Rowr turns
your Card Beame wheel and thus
two frising Mills Runs Verrey Easeiley
Altho Not So Good to take Care on for a frisear
Has them made in ye old way

to Cast a frising Board wich will Sound and
tough and Solid mix a pound of Vennis turpentine
in your Glew and Sand and Boyll itt well up and
Cast your Board About ye thickness of a Crown peice
and this Board i will maintain itt if properley
Dryed will Bee has Hard has Aney Stone 1773

Plate 1. Extract from the Brearley Memorandum Book, 1773. John Brearley was a prolific jotter of information picked up in the course of his work as a frizzing miller. He included many technical sketches, here noting a system of arranging gears in a mill. Source: Brotherton SC, YAS MS 2022, f.86 https://www.yas.org.uk/Collections/Brearley-Memo-Book.

technological upheavals during the eighteenth century. Its growing significance was well understood, but technical difficulties continued to plague its processes before 1790.[18] After that, new techniques adopted in smelting and forging iron transformed matters, and would help launch mechanical engineering as a trade in itself. Here, as in textiles, the 1790s marked the start of very significant long-term growth in manufacturing. The timing was not coincidental: the textile and iron trades moved forward in close association. In doing so, they involved many other settings, activities, and channels of knowledge.

Smail's view – of the textile merchants' mid-century marketing shift towards *scope* and the demands of fashion, rather than the narrower concentration on increasing *scale* – confirms that adaptability and an outward perspective were not born in the 1790s. That applies to the process of change more generally, in textiles, iron, commerce, and other pursuits. In many such endeavours, merchanting proved itself a source of dynamism.

But what of the regional industries which experienced commercial failure, undoubtedly declining over the course of the century? Where merchants in successful trades apparently became detached from wider developments, that could be attributed to a lack of engagement with knowledge flows. The textile examples are well known. A case can be made that merchants in East Anglia and the West of England failed their producers, the clothiers and manufacturers, in presiding over the eclipse of the local industry. They could perhaps be blamed for failing to pick up on the dangers of (for Norfolk) not securing worsted yarn supplies or (for Bradford-on-Avon) failing to invest promptly in a canal to enable steam power to work locally.[19]

The argument does not entirely stack up. East Anglia had more fundamental problems in maintaining a textile industry, with a niche product, lack of natural resource, and unwieldy organization. Smail sees manufacturers there as being 'secure in their niche, but also stuck there'.[20] But the Norwich merchants directing trade in that region had limited options, and nowhere else to go. The West's textile manufacturers did survive their change of circumstance, and thrived, though modestly. In contrast, Leeds merchanting underwent something of an upheaval, evident from the 1740s, but one to which some of the grandest merchant houses did not subscribe. Left to them alone, the town may not have enjoyed such robust good fortune later in the century. In the northern iron

---

18   See ch. 2.
19   Jenkins and Ponting, *British Wool Textile Industry*, 54–6, 71–2; Rogers, *Warp and Weft*, 71, 76–7; Mann, *Cloth Industry*, 154–6, 190. A Boulton & Watt engine was supplied to Wiltshire, to the Kennet and Avon Canal in 1796; Somerset had many Newcomen-style engines before 1800, mainly working in collieries. Source: John Kanefsky, presentation to the International Early Engines Conference at Elsecar Ironworks, May 2017; see also Dr Kanefsky's Early Engine Database, https://coalpitheath.org.uk/engines/index.
20   Smail, *Merchants, Markets and Manufacture*, 128.

trade, too, that period, the 1740s, was marked by the start of a shake-up in relationships, a fresh understanding of what constituted a merchant and an ironmaster. This, then, is not to be seen in black and white, as one region's merchants thrusting ahead while another's languished. It was more subtle.[21]

Transatlantic possibilities were increasingly significant from the late seventeenth century. Just as Britain's trade with continental Europe expanded fast, the Atlantic economy too began to offer great prospects. There were, says Jones, 'spectacular new opportunities for trade' for a new generation of merchants to export manufactured goods and bring in low-cost raw materials – 'literally a world of opportunities'. But that world was volatile, risky, and especially so because merchants operated in partnerships, or as sole proprietors, so were financially exposed. Limited liability did not then exist.[22] What is more, knowledge about how the North American trade functioned, and about its requirements, was hard-won. Back along the chain of supply, textile merchants aiming to make that Atlantic connection must convince British manufacturers to embrace a new mindset, to identify and produce saleable novelties. Often, that meant trial and error. Yorkshire cloth merchants, as Smail characterizes them, aimed to satisfy the customer, but then, 'the last thing goods did was move effortlessly'. It remained a great challenge 'to get the right goods from Yorkshire to the colonies, in the right colours, in the right assortment, at the right price, and at the right time'.[23]

Selling into Europe was routine business for the Leeds woollen-cloth merchant establishment. The wealthy young men apprenticed to the town's best merchant houses, usually for around five years and at a cost of many hundreds of pounds, were introduced to continental practices as well as the workings of the Yorkshire industry, and there, in particular, to the intricacies of cloth-finishing. They were trained in accounts and sometimes passed a year or more abroad during their education in the trade, for a working knowledge of the German, Dutch, and Portuguese languages was essential.[24]

Yet it was a different group which initiated cloth exports into North America. Barely twenty miles west of Leeds, merchants in Halifax had long been pushing the physical boundaries of trade. Their main product was kerseys, a woollen fabric, usually coarse, and considered to be clothing for the masses. In the sixteenth and seventeenth centuries, the town's merchant group opened up

---

21 For cultural contrasts between the cloth industries of the West of England and Yorkshire, see Randall, *Before the Luddites*, esp. introduction.
22 Jones, *Merchants to Multinationals*, 21.
23 Smail, *Merchants, Markets and Manufacture*, 1–5, 128–32; Smail, 'Manufacturer or Artisan?', 808.
24 Wilson, *Gentlemen Merchants*, 23–6, 63–4.

new markets from Prussia to the Levant, around the Mediterranean – and then into the New World.[25]

What does this say about merchant knowledge? Recruits to the élite body of Leeds cloth merchanting were launched into a well-defined programme which introduced them to manufacturing and prepared them to engage in commerce at home and on the continent. Though a formal approach, it had to accommodate change, technological and otherwise. The eighteenth century, unsettled by almost constant warfare, presented shifting obstacles to those exporting into Europe. The Leeds woollen-cloth trade adapted well to change early in the century, accepting (after some initial hesitation) several well-to-do Protestant cloth merchants fleeing continental persecution. That infusion of new blood partly accounts for a rise in the number of merchant houses, from an estimated 30 in 1700 to *c.* 50 by 1720. The Leeds trade subsequently flourished. Some long-established names withdrew over the course of that century, apparently a natural turnover. There was no shortage of new faces vying to take their place.[26]

The most notable entrants into Leeds merchanting, via an unconventional route which by-passed the usual terms of entry, were members of the Elam family. Their background was in woollen-cloth manufacture, their subsequent success built on cultivating business with fellow Quakers in Philadelphia. In the 1740s and 1750s, the father, Gervase, who had moved in 1703 from Halifax to Armley, close to Leeds, shipped small quantities of cloth across the Atlantic, while his sons John and Emmanuel set up as tobacco wholesalers, importing directly from America into Yorkshire. Knowledgeable about American commercial practices, adaptable to new circumstances, prepared to trade in other kinds of goods – all these qualities set the Elams apart from the main body of Leeds cloth merchants. In particular, though, they measured risk and embraced it, less daunted by the long credit terms that North American buyers expected. Their great success confirms that the family managed that essential balance, quickly responding to specific and complicated demands while maintaining competitive prices.[27]

Wilson credits the Elams with having 'largely pioneered the Leeds-American [woollen-cloth] trade'. Before 1750, Leeds had barely engaged with the Atlantic trade in cloth. But, as there was then no direct navigable route from Halifax to the Humber, Halifax merchants, at the forefront of opening new export markets, had to rely on Leeds facilities for onward packing and transit of cloth.[28] That physical barrier, and the greater market in Leeds for their tobacco imports,

---

25   Newton, 'Wakefield, Its Woollen-Cloth Trade and Merchant Networks', 136–7.
26   Wilson, *Gentlemen Merchants*, 17–23, and App. B, 241–8.
27   See Wilson, *Gentlemen Merchants*, ch. 3, esp. 49–50; on risk, 78–81; Smail, *Merchants, Markets and Manufacture*, 80–4; N.C. Neill, 'Gervase Elam (1681–1771)', DNB.
28   Wilson, *Gentlemen Merchants*, 32, 243, 58–9.

most likely determined the Elams' switch of focus. Then, once in business as woollen-cloth exporters, they exposed the complacency of those older merchant houses still holding firm to European markets, who saw America as too uncertain. The world changed around the old guard. Within a few years, the cloth frizzer Brearley noted a surge in the importance of American markets to 'trade from Leeds & Wakefield':

> London is the chief place and to Holland and Portugal and America Abundance of coarse goods goes to likewise to Russia and Hungary and Scotland and some small trade to Ireland But America is the cheiffest place for takeing the largest quantitys of goods in a general way. July 25th 1772.[29]

And this, 'goods in a general way', is significant. The boom which involved British manufacturers of textiles and iron goods was founded in long-distance trading which dealt also in slaves, sugar, hardwoods, tobacco, and many other commodities. Abraham Darby's cooking pots, the cloth from northern England – these were just part of a mass of merchandise feeding demand in the American colonies. Settlers and plantation owners wanted household goods, agricultural tools, and – as Angerstein chillingly but nonchalantly noted in Up Holland (Lancashire) in the mid-1750s – 'chains for slaves, sold from 3½d to 3¾d per pound'.[30]

Textiles and ironworking, the stand-out industries then and afterwards, were strikingly different in how they organized. The iron trade came of age in the Commonwealth period. Its technological needs – mining, importing and transporting, cultivating charcoal supplies, smelting and forging using a skilled workforce across multiple sites – demanded high levels of capital. Not only that, but it called for system in investment, in reducing risk by managing the various works as their fortunes rose and fell. This called for a distinct set of merchanting skills and knowledge flows.

The remarkable career of Ambrose Crowley (1658–1713), coinciding with that formative period in iron, illustrates the contrasts between his own trade and textiles. Early-modern cloth-making and marketing systems, like that centred upon Wakefield, had grown organically through a gradual spread of connection. The rise of Crowley and his iron-trade associates was spectacular in comparison, a series of strategic moves to build a sales and production conglomerate of national spread and international reach. The ambitious

---

29 Brotherton SC, YAS MS2022, f. 12.
30 Jones, *Merchants to Multinationals*, 17–21; Berg and Berg (eds), *Angerstein's Illustrated Travel Diary*, 303. It is not to be assumed that Quakers in this period were abolitionist. There was ambivalence, and some belief that slavery was necessary to economic development. See p. 165.

Crowley and his network show what was then possible, the benefits of influential personal connections and the extent of essential merchant knowledge.

Crowley was a merchant-manufacturer, a wholesaler of iron and steel products. His father, also Ambrose (c. 1635–1720), brought him up in the iron trade in Stourbridge and (outliving his son) continued to be a source of technical advice on iron and steel production. The older Ambrose had progressed from nail-maker to ironmaster, and by 1682 to steelmaker, prospering from opportunities arising after the seizure of Royalist-owned ironworks during the Commonwealth era. At one point he was a partner in Pontypool Forge with John Hanbury, a pioneer in manufacturing tin-plate. His son focused on the potential to trade with the royal dockyards, basing himself in a Greenwich warehouse. London, centre of the European iron market, was the place to form essential political connections. Stourbridge, though a hub of water transport for the Black Country, was difficult to access, via Bristol and the Severn.[31] Crowley built a vast integrated iron-production operation on Tyneside to supply escalating demands for nails, anchors, and other shipbuilding components. The Swedish iron that Crowley's operations devoured in bulk was routinely landed first at his wharf on the Thames, before being turned back northwards.[32]

North-east England had not been known for ironworking. After a problematic first venture in Sunderland in the early 1680s, when the arrival of specialists from Liège to train the workforce in Dutch nail-making methods fired up anti-Catholic violence, Crowley settled his manufacturing operations upstream of Newcastle upon Tyne. He opened sites on the lower reaches of a tributary, the Derwent, from 1691 at Winlaton, and later also at Swalwell, c. 1707–11. Crowley's new location attracted other expertise in iron- and steel-making. Denis Hayford, a south Yorkshire-based associate of the Spencer iron syndicates, founded the Company in the North to establish a steelworks at Blackhall, apparently a few years before Crowley arrived on Tyneside. From this, an influential cluster of iron and steel production grew in the Derwent valley, including the German Wilhelm Bertram, who became an innovative steelmaker at Blackhall.[33] Hayford was an especially valuable connection,

---

31 See Porteous, *Canal Ports*, ch. 5.
32 Price, 'Sir Ambrose Crowley (1658–1713)', DNB; Gerhold, 'Steel Industry', 79–83; Evans, Jackson, and Rydén, 'Baltic Iron and the British Iron Industry', 645; Barraclough, 'Development of the early steel-making processes', 110–12; also Flinn, *Men of Iron*; Hayman, 'Shropshire wrought-iron industry', 44.
33 Gerhold, 'Steel Industry', 79–83; Price, 'Sir Ambrose Crowley (1658–1713)', DNB; Cookson (ed.), *Victoria County History*, V, 76; Bowman, 'Iron and steel industries of the Derwent valley', esp. 85–100; Barraclough, 'Development of the early steel-making processes', 110–12, 115–16; Berg and Berg (eds), *Angerstein's Illustrated Travel Diary*, 258–72; Awty, 'Denis Hayford', DNB; Cookson, 'Wortley Forge', 54–7. For the Spencer syndicates, see ch. 6.

as his commercial interests spread across northern England, in Cheshire, Cumberland, and, in particular, south Yorkshire.

The older Ambrose had a more curious web of associates. Among his west Midlands circle was Andrew Yarranton, with whom he reportedly visited Saxony in 1667 to see how tin-plate was produced.[34] His son must also have known Yarranton, and Dud Dudley, though those two were not on good terms. Both were published authors. The Oxford-educated Dudley had written about his attempts to smelt iron with mineral fuel – coal rather than charcoal.[35] Yarranton (1619–84), forward looking and remarkably wide-ranging, was a civil engineer, fought as a captain in the parliamentary army in the English Civil War and became a proto-political economist whose ideas were still received seriously two centuries later.[36] His economic prospectus was set out in *England's Improvement by Sea and Land to Out-do the Dutch without Fighting* (1677–81). The second element of the title was dropped for the book's posthumous reissue in 1698, after William of Orange acceded to the British throne as William III.

But Yarranton had flattered the Dutch and Germans by seeking to imitate their most successful ventures. He suggested introducing a registry of deeds, something that already 'advanceth trade in Holland', against which loans could be secured as mortgages.

> A Register will quicken Trade, and the Land Registered will be equal as Cash in a mans hands, and the Credit thereof will go and do in Trade what Ready Moneys now doth.[37]

Yarranton also used successful German and Dutch models to advocate river navigation schemes – deepening and improving rivers, specifically the Humber.[38] His ideas were picked up around the turn of the century with the Aire and Calder Navigation and the West Riding Deeds Registry, both highly influential in Yorkshire's subsequent industrial development in iron as well as textiles.[39]

---

34   Howes, 'Why innovation accelerated', 28–9. The Saxony method was a forebear of what Angerstein witnessed much later in Pontypool: see p. 32.
35   See p. 40.
36   E. Clarke, rev. P.W. King, 'Andrew Yarranton', DNB; P.W. King, 'Andrew Yarranton', BDCE; Smiles, *Industrial Biography*, 71–3. Yarranton was, for instance, quoted by the economist W.S. Jevons in 1865, in *The Coal Question: An Inquiry Concerning the Progress of the Nation, and the Probable Exhaustion of Our Coal-Mines*, ch. 5.
37   Yarranton, *England's Improvement*, 12. 'Out-do the Dutch' was removed from the book's title on its reissue in 1698, after Yarranton's death and the accession to the British throne of William of Orange, whose reign saw an upheaval in financial institutions which would underpin eighteenth-century economic developments.
38   Yarranton, *England's Improvement*, 4–5.
39   See Sheppard and Belcher, 'The Deeds Registries of Yorkshire and Middlesex'; Tate, 'Five English District Statutory Registries of Deeds'.

ENGLAND'S
Improvement
BY
SEA and LAND.
TO
Out-do the *Dutch* without Fighting,
TO
Pay Debts without Moneys,

To set at Work all the POOR of *England* with the Growth of our own Lands.
To prevent unnecessary SUITS in Law;
With the Benefit of a Voluntary REGISTER.
Directions where vast quantities of Timber are to be had for the Building of SHIPS;
With the Advantage of making the Great RIVERS of *England* Navigable.
RULES to prevent FIRES in *London*, and other Great CITIES;
With Directions how the several Companies of Handicraftsmen in *London* may always have cheap Bread and Drink.

By *ANDREW YARRANTON*, Gent.

LONDON,
Printed by R. *Everingham* for the Author, and are to be sold by T. *Parkhurst* at the Bible and three Crowns in *Cheap-side*, and N. *Simmons* at the Princes Arms in S. *Paul's* Church-yard, M DC LXXVII.

Plate 2. The title page of Andrew Yarranton's *England's Improvement by Sea and Land* (1677) illustrates the breadth of his industrial and economic vision.

The younger Crowley, exposed to such influences and with considerable flair of his own, was not a typical merchant. Despite his level of success, though, fundamentally that was what he remained: part of a wider world of merchanting, communicating with many ordinary people as well as some who were extraordinary. His career reflects that broadness of scope, as he pushed at the boundaries of what was then available and possible.

Knowledge was key to merchanting. Information was strength, and in these two key industries, iron and textiles, its consistent turnover is plain to see.

Eighteenth-century merchants built from systems and bodies of knowledge which were long-standing. The iron trade experienced an obvious disjunction (in the mid-seventeenth century), while growth in textiles was more organic at that point. But there are very many similarities between the progression of these two activities, most notably national and international knowledge movements through transacting with continental Europe and, increasingly, with North America.

British iron's strong – and long – relationship with Sweden gave access to Swedish expertise in ironmaking as well as supplies of iron and steel. Crowley's experiences indicate the texture of relationships with other overseas interests in the earlier period: his father's trip to Saxony to learn about tin-plate production; Yarranton's ideas about infrastructural developments gleaned from Germany and the Netherlands, as well as his wider economic view. They show specialist skills imported from Liège to a British workforce, and the arrival of Bertram and others from Germany to make steel on Tyneside. Later, there was Angerstein, who, it emerges, was only one in a century-long list of official Swedish visitors engaging with British iron interests in order to understand and improve processes and products.[40]

Branches of the textile trade were also long attuned to continental suppliers and customers, with their own useful conduits for technical and commercial data. The many-tentacled Wakefield textile production chain relied on customer feedback to develop new products, even though some of that came at second hand, via the London cloth hall. Defoe reported that 'many travellers and gentlemen have come over from Hamburgh, nay, even from Leipsick in Saxony' on purpose to see the busy villages feeding Leeds's woollen-cloth markets.[41] Foreign textile workers, including merchants, brought valuable skills to the British trade. The most noticed were Protestant refugees, Huguenots such as Lewis Paul, who stood out with his notably innovative ideas. And, as the Elam family illustrated, it was possible to overturn custom and received wisdom, to risk and flourish.

The shake-up in Leeds after 1750 is just one example of merchants reacting and adapting to new circumstances. Change was intensifying, and across many industries and regions. The Elams were not alone in rethinking their own merchanting model, in understanding commerce as an activity in itself, involving themselves in more than one type of product, acquiring knowledge, and adapting it to new purpose. Others looked further into foreign markets, gathered specialist experience and became enablers, transactors, stimulators of products, and organizers of process. This could be organized and explicit, while other interventions were reactive and off-the-cuff. A number

---

40 See ch. 2.
41 Defoe, *Tour thro' the Whole Island*, Letter 8, Part 4.

of the dealers who joined the Leeds cloth trade from the mid-century had no background in woollen manufacture or cloth-finishing. The new men instead offered financial substance, or fresh perspectives. Among the most effective were specialists in other kinds of merchanting: wholesale grocers, tobacconists, drysalters, textile retailers. As cloth merchants they went on to earn respect for their part in the Yorkshire trade's expansion, and for their integrity and acumen.[42] Responses in other regions differed according to local conditions. In Manchester, as in Leeds, foreign merchants came in and proved significant in developing industry, while in the port of Liverpool, specialist commercial structures for cotton goods emerged.[43]

The technical progression in ironworking, from the late seventeenth century, is explored at length in the next chapter. Regarding merchant knowledge, the way in which nail-making spawned influential forge-masters and nail merchants is notable. The Crowleys are examples, and there were many others. To become an organizer of the trade offered far more comfortable prospects than the sparse returns and drudgery of domestic nail-making, which was generally fitted in around agriculture's seasonal demands. Nail merchants' profits bought partnerships in ironworks, where they shared their own commercial and managerial experience. Other nailers found ways into forge ownership. Connection was reinforced, new pathways for knowledge created. By the century's end, some of these men brought up in nail-making redirected their manual aptitude into innovative ventures in metal-working and mechanical engineering. For those in intermediary roles, too, opportunities opened up – for agent-managers with expert knowledge of mining and working minerals, and specialist clerks in commercial settings. Knowledge was key, or becoming so.

The context of knowledge

Knowledge was key, but key to what? Its utility – and hence its value – might be highly specific.

These industries, and the commercial activity associated with them, contained multiple strands of expertise. The merchant examples show some of the variety of specialisms at work, within a context which can appear confusing. For a perspective on this, on the framework of groups and dynamics, it is helpful to think in terms of communities. In defining 'community', the many gradations of social class have also to be considered – because the social and industrial background was responsible for delivering certain people into certain occupations.

---

42 Wilson, *Gentlemen Merchants*, 32–3.
43 Jones, *Merchants to Multinationals*, 22.

'Community' is a collective experience. It could be, and often was, something existing outside a geographical locality. A community might consist of groups of people within one industry, or people who shared an expertise, an education, an academic endeavour, a belief system. It could be an extended family, perhaps intermarried, maybe migrants with some interest in mutual advancement and support. It may be localized, though communities can reach beyond regions or nations. But certainly 'community' had agency in the innovation process, and could be many things.

Also to be underlined is the degree to which individuals were identified with their social standing, at a time when society was finely graded by class, and class within class. Artisans, the skilled manual workers whose role in industrialization was pivotal, were closely associated with their own trade, often one practised by previous generations and other family members. This had practical significance: being part of a trade community, working with a degree of mutuality, offered ways to alleviate threats and uncertainties. The personal relationships are not to be idealized, for very many instances of business and family dealings show individuals in conflict, and conduct that can only be seen as undermining others' interests. People knew this, but still the fragility of eighteenth-century life and assets was a strong motive to seek protection against personal disaster. What is more, the emerging technologies themselves – and specifically their limitations – meant that cross-trade coalitions and other kinds of co-working became a sensible, if not inevitable, contingency measure. That was a reality in many working situations for artisans and small businesses. Cross-fertilization – the meetings of minds, experience, and knowledge – had the potential for positive outcomes, in cultivating skills and abilities into something novel. Industrialization was actively pursued, not passively received.

And of course, classes did connect and collaborate, in the workplace and elsewhere. Class was very real, but it was not absolute or immutable, and, while it related to wealth, it was not measured solely in monetary terms. Contemporaries would have intuitively grasped the intricacies, at least as these affected them personally. Nor did inherited expectation affect only the most privileged. Social class determined pathways into artisan and commercial trades, with sons following fathers through apprenticeship into what was seen as a family occupation. Commonly, a lack of means, education, and connection held back individuals from fully achieving their potential. Yet a higher social standing could deter any engagement in a manual trade, just as eyebrows were raised when the wealthy Leeds merchant Benjamin Gott turned to manufacturing, albeit at arms' length.[44]

Class is better understood as a spectrum rather than a separation. E.P. Thompson saw it as a relationship, not a thing in itself: 'Class is defined by men

---

44 Class in these industrial circumstances is further defined in chapter 4.

[sic] as they live their own history, and, in the end, this is its only definition.' It is embodied in real human beings and in a real context. Above all, it is *historical*, and needs to be discussed in a time and place. So people and events are not to be seen through a prism of theory, or through ways of relating which developed in later generations. For, as Thompson argues, the danger is that we 'obscure the agency of working people, the degree to which they contributed by conscious efforts, to the making of history'.[45] Particularly in industrial communities at the forefront of early technological change, to focus only on higher social ranks means missing much of the dynamic.

The spread of knowledge

What has already been described – and is to be further elaborated in the following chapters – suggests that the fundamental change afoot by the mid-eighteenth century drew on channels of knowledge which conveyed the energy and experience of many regions and places. Evidence reveals significant commercial transaction between British regions. Industries connected, they innovated – and this process was intensifying across the nation. Change was already becoming embedded: in products, materials, techniques, transportation, and in systems and methods of working. It was driven on, too, by communities repurposing to meet the new needs, and by their increasingly outward-looking perspective on the world.

What, then, is to be made of Joel Mokyr's argument that in 1800 much of the nation lived in ignorance of the major upheaval then in progress? If true, this lack of awareness would have significantly inhibited innovation and economic growth – for instance, by holding back the adoption of new technologies, or not encouraging the migration necessary for expansion. Is it possible that localities and communities, even sizeable parts of regions, remained untouched by, possibly wholly ignorant of, recent developments? Based on the evidence here, this seems far-fetched.

Mokyr suggests that 'daily material life' changed little 'outside a few key areas and regions ... and the average Briton in 1800 was probably only dimly if at all aware that something very large was brewing on the horizon'. Before 1800, 'most of Britain's industry was still located outside the modern sector and hardly experienced any technological change', he says.[46] Perhaps this argument rests on a closely defined 'technology', of certain famed innovations in certain key industries. Even so, the weight of evidence suggests otherwise. Overwhelmingly

---

45 Thompson, *Making of the English Working Class*, preface.
46 Mokyr, *Enlightened Economy*, ch. 5, esp. 79, 95.

it points to technological advance, in all its forms, as a widespread phenomenon and one widely engaged with.

'What was brewing' implies that this was about forecasting the future, something that we know but that contemporaries could not. But prediction is not the issue. In 1800 it was already clear that 'something very large' had been happening for some long time. Change was already prevalent, endemic in the decades leading to 1800, and across very much of the nation. There may be interesting questions to ask about the extent to which people in general saw the bigger picture, the major changes that were happening around them and further afield; and how they responded to these new circumstances. But by the latter stages of the eighteenth century, industrial change, accompanied by urban growth, was highly visible. Not all localities adopted the latest discoveries, but that is a different matter from being unaware. Technologies were not selected for their novelty, but because they suited local needs and resources.

Across the land, wherever there were mines or textile manufacturing or ironworks, there were steam engines; water-powered factories (more than 200 Arkwright-style cotton spinners ranging across the Pennine countryside by 1794) and mills to prepare wool for spinning; basic new domestic technologies in thousands of cottages and domestic workshops by the 1770s; canals and waggon-ways connecting multiple localities; and docks, cranes, and bridges. Such installations were everywhere, in rural as well as urban settings. Indeed, most watermills, and many miles of canal, were by necessity rural. More generally, while collieries, textiles, and metal-working industries across northern England appeared especially open to change, those pursuits and places had no monopoly on innovation. Novel products, processes, and machinery were evident across many other regions, notably in south Wales, the English Midlands and West Country, lowland Scotland, and London, as we will see.

Knowledge moved where people moved, and people moved a lot. There were no walls around localities, and mobility was far more common than we would perhaps assume. That fluidity, whether temporary or longer term, disseminated skills between regions and within them. Networks, personal relationships, spread the word to inform and encourage migration, bringing aptitudes to where they were needed. In consequence, Manchester, the Black Country, Bristol, Tyneside, Leeds, Shrewsbury, and many other places grew, all in their own fashion, and fired by information that had spread out beyond their boundaries.

For the newcomers, and for the towns, this was a reciprocal process. Building a concentration of aptitudes and perspectives had direct impact on industrial development. With this came opportunities for a meeting of minds and experiences, and a sharing of knowledge which nurtured individual skills and abilities. The product could be something very novel.

Waves of change had lapped around Britain's industrial activity for some while, even before 1700, and began to crash with greater intensity afterwards.

Momentum was building. One core question relates to these channels of experience and learning: when, where, and how did they become a route of knowledge which ultimately fed innovation?

The answers come in empirical form, corroborated by exploring the eighteenth-century industrial world and its context. Much of what was most significant is not really measurable, because it comes down to relationships – such as that between knowledge and innovation – and above all links between people. But those connections can be identified, defined, evaluated, seen in real time, and their consequence demonstrated. The substance of those connections is not to be assumed.

2

Innovations in iron

Angerstein's industrial picture

In 1753–5 a Swedish visitor, an intelligent stranger of unbounded curiosity, travelled through the industrial districts of England and Wales. Reinhold Rücker Angerstein kept detailed illustrated journals, from which emerges an exceptional record of a country in the process of transformation. Angerstein came well equipped for this mission, profoundly knowledgeable about iron-trading and ironworking in his native land. He was also well positioned to contrast what he saw in Britain with his own observations of industrial sites on the continent, which he had toured extensively over the previous four years.[1]

Born in 1718 to a prosperous ironmaster with German antecedents, Angerstein was a university-educated and highly valued Swedish government employee. He continued in this official position alongside running the family ironworks after his father's death in 1734. The tour of Europe took him away from home for six years, from 1749. Angerstein bore most of the cost himself, though the Swedish Ironmasters' Association and the century-old *Bergscollegium*, the Board of Mines which oversaw mineral workings, gave instructions and some financial support. He toured England and Wales – though not Scotland – from September 1753 until early in 1755.[2]

On returning to Sweden, Angerstein intended to apply some of his discoveries in the family business, including a new foundry with a reverberatory (air) furnace, modelled on one seen in Britain. To that end, he acquired the Vira sword works in 1757, planning to reorganize the old blade and scythe smithy. The scheme was strongly resisted by his smiths. Confident, well informed and pragmatic, and with

---

1   Before Berg's translation, published by the Science Museum in 2001, the diaries were available only in Swedish, as a fair copy transcribed after Angerstein's death and kept at the Swedish Ironmasters' Association Library, Stockholm. Some of his original working documents are in the Swedish State Archives: Berg and Berg (eds), *Angerstein's Illustrated Travel Diary*, xix. I am much indebted to Göran Rydén for comments and suggestions on an earlier draft of this chapter; and to John Suter for technological advice.
2   Berg and Berg (eds), *Angerstein's Illustrated Travel Diary*, xviii; Rydén, *Production and Work*, 7.

official backing, Angerstein expected to convince them to fall into line. But he became ill, and died in 1760 before anything could be accomplished.[3]

Angerstein's tour coincided with a pivotal period in relations between the Swedish and British iron trades. Sweden was concerned about possible threats to its position as Britain's main supplier of bar-iron. Angerstein had been charged with assessing the use of iron from other sources, notably Russia and North America, and also possible threats to the Swedish trade from British technological innovations.[4] This adds to the fascination with Angerstein's accounts of what he witnessed in England and Wales. Yet it was only one focus of his trip, and Sweden would remain Britain's main source of bar-iron into the 1760s.[5]

Nor was Angerstein the only expert Swedish visitor, though his fine sketches and a text now available in translation have made him the best known. A steady stream of similarly investigative Swedes, sometimes (and often quite mistakenly) referred to as spies, is evident from 1692 and into the mid-nineteenth century. Angerstein's contemporary Samuel Schröder (1720–79), visitor to England in 1749, is considered the better observer and writer, though without illustrations, and had a higher profile in Sweden. Sven Rinman (1720–92), the greatest Swedish metallurgist of his time, was another noted traveller, although not to Britain. Rinman's reputation grew from understanding the practicalities of the trade while also publishing extensively on science. After Angerstein died, his factory continued to supply the state with swords, drawing on technical advice from Schröder and Rinman.[6] The Swedes, as trading partners, evidently met a warmer reception in Britain than, for instance, the French metallurgist Gabriel Jars, who arrived in the 1760s with a highly specific brief to investigate the iron and steel industries.[7]

While his expertise lay in ironworking, Angerstein took in a wider view of industrial practice. His notes, a snapshot of early-1750s British industry, reveal widespread changes in progress in manufacturing, as well as mining, colliery railroads, and the processing of various metals. He was interested in steam

---

3 Berg and Berg (eds), *Angerstein's Illustrated Travel Diary*, xvi–xix; Evans and Rydén, *Baltic Iron*, 121–4 and *passim*.
4 Berg and Berg (eds), *Angerstein's Illustrated Travel Diary*, ix, introduction by Marilyn Palmer.
5 For Angerstein, Rinman and other Swedish specialists, and a new overview of technical development there, see M. Jannson and G. Rydén, 'The Oeconomia of Iron and Steel: Material Transformations, Manual Skills and Technical Improvement in Early-Modern Sweden', in Nigro (ed.), *Knowledge Economy*, 237–62.
6 Evans and Rydén, *Baltic Iron*, 13, 15, 54, 122–4, 310 and *passim*; Jansson and Rydén, 'Improving Swedish Steelmaking'. Göran Rydén advised on the status of Angerstein, Schröder and Rinman in Sweden. See also Harris, *Industrial Espionage*, esp. 509–15, 291–2. The DNB itself contains an article headed 'Industrial spies', which unhelpfully fans this flame.
7 Harris, *Industrial Espionage*, 224–36; Evans and Rydén, *Baltic Iron*, 143–6; see also Jars's own posthumous publication, *Voyages Métallurgiques*.

power and other prime movers, and gathered evidence of the costs and possibilities of smelting iron with charcoal or coke. He paid close attention to how production was organized to best economic effect. He observed, he sketched, he interrogated. He noted equipment and tools, technique, organization, wages, transport, and other costs, and the views and experiences of companies purchasing iron. Sometimes he fell under suspicion and was quickly dismissed. Less than twenty-four hours after arriving in Sheffield, he was advised to leave town for showing too much interest in Huntsman's crucible-steel works. But in the main, he was welcomed and accepted. After all, he represented an important supplier, and readily shared useful information with those he called on.[8]

On arriving in Britain, Angerstein was perfectly primed to identify affinities and differences between Swedish, British, and other European practices. He spotted, for instance, that the Aston blast furnace was built like ones in Hanover and elsewhere in Germany. He noted that in Birmingham files were cut exactly as in Steiermark, Austria, and Schmalkalden, Germany, but that here they were ground before cutting. When ordered off Hanbury's premises in Pontypool by the proprietor, Angerstein was able to dispute the point: secrecy was futile, as Hanbury's methods had come to England from Liège and were already widely disseminated. The traveller conceded, though, that the Pontypool rolling mills were a considerable advance on what he had seen in Liège and Namur, 'where the rolling of sheets is unknown and not even considered possible'.[9]

Angerstein was interested in the uses to which different grades of iron were put. Bar-iron was not homogeneous. Öregrund bar-iron, from eastern Sweden, was the world's finest, and best for steel-making. Its name came from the Baltic port through which it was exported from the *Vallonbruk* in Uppland, a group of fewer than twenty ironworks practising the Walloon method of forging. This technology had been introduced by migrants from the Netherlands in the seventeenth century. Öregrund iron was prized in England by the Royal Navy, and for converting to blister steel. Barnsley wire-drawers used it, although, as the price was about 25 per cent higher than English iron, only to make needles for Nottingham stocking-frames. Wire for carding wool used the cheaper English product. Generally, Swedish iron was stronger, harder, and more resilient than any other, and it was used for a wide range of goods

---

8 Berg and Berg (eds), *Angerstein's Illustrated Travel Diary*, xvi; Rydén, *Production and Work*, 7–10.
9 Berg and Berg (eds), *Angerstein's Illustrated Travel Diary*, 33–6, 163. Hanbury's works, 1695–7, introduced a water-powered rolling mill with adjustable rolls, a major step in enabling the establishment of a tin-plate industry. In *c.* the early 1720s the company overcame the problem of cleaning the blackplate prior to tinning, and was the first in Britain to produce tin-plate on a commercial scale: P. Riden, 'John Hanbury, 1664?–1734, landowner and ironmaster', DNB.

including horseshoes, nails, and anchors. The softer English iron went for rolling into plates, and also for nails.[10]

So Angerstein's was a practical, comparative, observational approach. His compatriot Schröder described more explicitly what the Swedish trade might learn from Britain, such as the 'English method' of organization and division of labour. Angerstein's journals concentrated on recording and reviewing points of excellence and originality. Darby's pig iron at Coalbrookdale, coke-smelted and used for casting, particularly impressed him. Charcoal cost ten or fifteen times more in England than in Sweden, and production costs of English bar-iron were double those in Sweden. Still, charcoal-smelted iron was cheaper than that made with coke, which could not then be used to forge bar-iron. This was the problem finally solved by the puddling process. Angerstein admired English foundry techniques, notably at Coalbrookdale; and also the cast hammer-heads and anvils used in English bar-iron forges, a type not then replicable in Sweden. Like Schröder, he was struck by British production systems, especially how iron fabrication was organized in Birmingham. Parts moved through a series of workshops – each worker specialized in one task, so that speed and quality were increased by repetition, and costs cut. Ambrose Crowley's empire of large ironware factories on Tyneside and in the Derwent Valley also proved of great interest, as did Bertram's nearby steelworks.[11]

Angerstein set a scene, and at a critical point. Yet there is something significant that was not said, a frame of reference that was missing. An example is given in his summary of the south Yorkshire ironworking centres of Barnsley and Wortley. Angerstein discussed the iron available, and the range of opinion among wire-drawers in their preferences. Some claimed superiority for local products. Others described the qualities they needed, and thought they recognized these in a particular type of iron. Yet none was quite able to explain why one iron was better than another: 'It is not possible to find anybody who really understands it and can explain the difference in quality between ... wire drawn of Öregrund iron, and English wire made of English iron.' Some said English wire was softer, others considered it to be hard, but evidently each satisfied the workers in iron, in many quite different applications, in a way which to Angerstein was not entirely logical.[12]

10   Evans and Rydén, *Baltic Iron*, 65–6, 79, 320, 323; Berg and Berg (eds), *Angerstein's Illustrated Travel Diary*, 214–16, 218, xviii.
11   Evans and Rydén, *Baltic Iron*, 13, 15, 54; Berg and Berg (eds), *Angerstein's Illustrated Travel Diary*, xvii, 329–38, 258–67; Barraclough, 'Development of the early steel-making processes', 110–12, 115–16; Berg, *Age of Manufactures*, 255–7. See also Rydén, *Production and Work*, 29, on costs of charcoal and coke.
12   Berg and Berg (eds), *Angerstein's Illustrated Travel Diary*, 218; Berg, *Age of Manufactures*, 257.

But in considering this enigma, Angerstein himself fell back upon a few basic adjectives in order to define the features of various types of iron. This man, an authority on the European iron trade, suggested nothing of a science of metallurgy, or any fundamental principles, chemical or physical, of working with iron, to illustrate or explain his point. This is not entirely surprising, for leading British ironmasters, even as late as 1800, showed little or no interest in learning more about the basic scientific processes underpinning their activity.[13] Yet other Swedes who wrote about iron, notably Rinman, attempted a scientific rigour when questioning the differences between iron types. Angerstein's evident lack of engagement with science, as it then stood, is to be underlined.[14]

Britain, Sweden and the iron trade

Angerstein's observations show Britain as a hive of innovation, with inventiveness widely spread across many industries and regions. Creativity in manufacturing encompassed new sorts of product, and novel devices and techniques to make them. Considerable resourcefulness was on display in adjusting production systems, to satisfy increased demand and to improve quality and efficiency. Individual businesses did not necessarily address such problems – often or at all – but out of collective endeavour came countless useful advances, over a wide spectrum of significance. Angerstein captured a snapshot of slow but meaningful transition, a continuation of craft and domestic production alongside the adoption of significant innovations, and budding regional specialisms. He witnessed Newcomen's atmospheric engines in use, Darby applying coke to iron-smelting, silk-throwing powered by water, and Kay's flying shuttle at work in weaving woollens. In the early 1750s he was too early for Watt's steam-engine improvements, Cort's puddling process, or Arkwright's water-powered cotton machinery.[15]

As for the iron industry which was Angerstein's primary concern, historians have disagreed about its state of health around that time. Its forge sector then depended almost entirely upon charcoal-smelted bar-iron, much of this imported from Sweden or Russia. By the 1780s, Baltic imports satisfied two-thirds of British demand, so twice as much as home production.[16] This does not mean that the domestic industry was in the doldrums: Philip Riden

---

13 Cookson, *Age of Machinery*, 177.
14 See p. 72.
15 Berg and Berg (eds), *Angerstein's Illustrated Travel Diary*, ix–x, introduction by Palmer. For a comprehensive and long-term review of British iron technology, particularly forging, see Hayman, 'Shropshire wrought-iron industry', esp. ch. 2.
16 Evans, Jackson, and Rydén, 'Baltic Iron and the British Iron Industry', 662; Evans and Rydén, *Baltic Iron*, 13–18, covers something of the historiography.

established that the production of British charcoal-smelted iron in 1750 was probably greater than it had ever been. Riden also challenges the idea that this industry had ever been troubled by fuel shortages, or experienced decline. The era of charcoal blast furnaces started in c. 1540, and the last of that type were built around 1750.[17] Certainly Angerstein found charcoal in widespread use for smelting, and although some complained of shortages, caused partly by poor husbandry, many ironmasters and their charcoal suppliers were evidently careful managers of woodland. He witnessed coke-smelting only at Clifton (Cumberland), and at Darby's Coalbrookdale, where the iron thus produced was unsatisfactory for bar-iron and used only for castings.[18]

How problematic was the long-standing dependence on charcoal? Supplies may not have been under immediate threat, but any major expansion dependent on home-produced iron would have been unsustainable. It is said that the Old Furnace built by Thomas Brooke at Coalbrookdale in c. 1658 required 4,000 acres of woodland to fuel it.[19] How far were the looming difficulties anticipated before 1750? To contemporaries with any interest in this, it must have been obvious that the British iron industry was increasingly stretched geographically. A survey of English and Welsh furnaces and forges in 1717 shows concentrations in the Forest of Dean and south Wales; large numbers further north on the Severn, in the west Midlands, and into north Wales; and a declining group on the Weald. There was also a string from north of Derby to Leeds, centred on Sheffield, and a handful in north Lancashire.[20] For charcoal-fired smelting, blast furnaces needed close proximity to extensive woodlands, and to both water power and water transport. From 1720, most new furnaces were coastal, in north Lancashire or Cumberland, far north of the west Midlands where the bulk of demand for pig iron lay.[21]

Almost universally, industries which were the greatest consumers of fuel adopted coal during the seventeenth century, or even earlier. Entirely new industrial centres became established in places where cheap coal was abundant. Around the port of Sunderland, for instance, otherwise unsaleable low-grade coal, which had fired salt pans in the late sixteenth century, was afterwards the making of local trades in glass, pottery, lime, and brewing. Iron production, in its persistence with organic rather than mineral fuel, was out of step.[22]

---

17 P. Riden, *Gazetteer of Charcoal-fired Blast Furnaces*, 6.
18 Berg and Berg (eds), *Angerstein's Illustrated Travel Diary*, ix, introduction by Palmer.
19 Inf. Ironbridge Gorge Museums. 4,000 acres is more than 16 sq. km.
20 Schubert, *History of the British Iron and Steel Industry*, 192–3.
21 Evans, Jackson, and Rydén, 'Baltic Iron and the British Iron Industry', 661.
22 Ashworth, *Industrial Revolution*, 169, 172–3; Evans, Jackson, and Rydén, 'Baltic Iron and the British Iron Industry', 662; Cookson (ed.), *Victoria County History of Co. Durham*, V, 75–6, 234–6.

As Britain came increasingly to dominate the international market in bar-iron, this trade became conspicuously outward-facing in comparison with many others. The stream of wars which disrupted eighteenth-century European markets, bringing with them uncertainty along with additional tariffs and taxation, also generated a brisk business in armaments and other iron products. The geo-politics had the further effect, of course, of amplifying long-running British anxieties about the nation's over-reliance on Swedish iron. Finding a substitute for Öregrund iron was seen as a priority, especially considering the Royal Navy's dependence on it.[23] After 1660, Swedish bar-iron had overtaken Spanish in its penetration of the British market. A diplomatic crisis in 1717–18 prompted the first of several attempts to reduce or eliminate this reliance on Sweden by promoting ironmaking in the British colonies. North American forests in particular seemed to offer boundless opportunities for charcoal. Ultimately the idea of a charcoal-powered colonial pig-iron industry faltered, defeated by politics and wrangles over import duties.[24] Was this unsuccessful scheme, in the crisp phrase of Evans and Rydén about Britain's unhealthy dependence on Sweden and Russia, 'anything less than a raid on the vegetable energy stocks' of other nations?[25]

Concerns about the situation were mirrored in Sweden. During the seventeenth century, the country had built Europe's largest export trade in bar-iron, virtually a global monopoly. From the mid-century, Britain overtook the Dutch as Sweden's largest customer. Almost half of all bar-iron exported from Sweden in 1700, 15,000 tons, was supplied to Britain. This market continued to rise, to around 20,000 tons annually through the mid-eighteenth century. Sweden, initially confident in the superior quality of its products, began to recognize Russian iron from the Urals as a competitive threat. Charcoal production was highly labour-intensive, and Russia's main advantage over Sweden came through its use of serf workforces in well-managed charcoal forests. In the late 1720s, Russian iron could be bought on the London market for £13 to £14 a ton, while Swedish products (though not directly comparable in quality) were selling at £17 5s., and up to £19 a ton. In the 1750s, the volume of Russian iron entering Britain matched that from Sweden, and then surpassed it.[26]

Evans and Rydén emphasize the iron *trade* as exactly that – a commercial web spanning continents, serving dispersed manufacturing specialisms, and

---

23   Ashworth, *Industrial Revolution*, 10; Rydén, *Production and Work*, 14–15; Evans, Jackson, and Rydén, 'Baltic Iron and the British Iron Industry', 646.
24   Evans, Jackson, and Rydén, 'Baltic Iron and the British Iron Industry', 659–61. The crisis of 1717–18 and the potential of pig iron from the colonies are described by Ashton, *Iron and Steel*, 110–14.
25   Evans and Rydén, *Baltic Iron*, 40.
26   Rydén, *Production and Work*, 13–14; 'Iron production and the household', 78–9; Evans, Jackson, and Rydén, 'Baltic Iron and the British Iron Industry', 656, 659.

crossing frontiers. The concept is called a global commodity chain. When Samuel Schröder identified an 'iron system' in 1749, this is what he too acknowledged. Policy-making and geo-politics were as necessary features of the system as the 'dealing and manufacturing' which Defoe, earlier, had understood by the term 'trade'. By the mid-century, Britain was already central to the iron trade, which was fluid and diversifying, while ironworking itself rested on a division of labour across national boundaries.[27] But while the 'iron system' had so far proved adaptable, as demand boomed and with vegetable resources stretched and ultimately finite, how sustainable could it be?

Ironmaking technologies

The chain's fragility was exposed and exacerbated by European warfare. British ambition faced some realities. For the sake of national security, more iron must be produced. Yet the days of self-sufficiency in charcoal were long past, and this was itself a threat in troubled times. The solution now seems obvious: a shift to mineral fuel, in which Britain was rich. This would require coal-fuelled ironworking technologies.

Why did it take so long? The obstacle preventing coke-smelted pig iron from being converted to bar-iron was decisively eradicated by the rolling and puddling method attributed to Henry Cort, patented in 1783–4 and available in practical terms from the 1790s. However, none of this, even the possibility of coal becoming so central to ironmaking, was fully apparent to contemporaries until Cort's process was actually made to work, so after 1790.[28] This point is a very significant one in explaining the course of ironworking technologies through the preceding century.

Cort's intervention, then, was critical, a great industrial discovery of the eighteenth century. He built upon procedures already being tried elsewhere, but according to Chris Evans, most of his contemporaries

> had no doubt of the genuine originality of his contribution. It was puddling and rolling, not any other of the rival techniques current in the 1780s, which remained the basis of iron-refining for decades afterwards, until wrought iron was superseded by mass-produced steel in the second half of the nineteenth century.[29]

---

27 Evans and Rydén, *Baltic Iron*, 12–15, 54, 230.
28 Evans and Rydén, *Baltic Iron*, 40, 250.
29 Chris Evans, 'Henry Cort (1741?–1800)', DNB.

Yet, as with most cases of overnight success, there is a long back story. And like many innovations, Cort's method took some time to be universally welcomed and adopted, even though his discoveries were effectively unpatented.

With the first introduction of blast furnaces, from the late fifteenth century, an indirect process replaced an older, one-stage method of reducing ore in a bloomer. In that era, ore was first smelted by charcoal in a blast furnace and then refined by forging. Before 1700, cast iron was produced by being poured into moulds direct from the blast furnace. With a carbon content of 3 to 4 per cent, it was too brittle to be worked by hammer. To make wrought iron – malleable iron – it needed further refining in a 'finery forge'. The final process, to make bar-iron, required hammering into shape in a chafery hearth. The vital drivers of iron production were water to power bellows and hammers, and a considerable input of manual skill and strength.[30] The challenge Cort addressed was to find a process which worked equally well in forging coke-smelted iron.

Surrounding Cort's innovation, and even more so that of Darby decades earlier, there are many questions. Why did success take so long to show? Why such delay in implementing breakthrough technologies? And another disconnect cries out for explanation: we are told there was a boom in British iron production after 1750, and yet a supposed stalling before 1783. How is this to be accounted for? Something else was going on. Are there technological explanations – other useful innovations by people less known, or supplementary fixes which had great impact?

There is indeed a wider picture, and a story extending back into the seventeenth century. The reverberatory (air) furnace, in which pig iron is refined in a chamber separate from the fuel, preceded Darby. So did an associated proto-puddling technique described by Plot in the 1680s. The considerable level of expertise employed in contemporary iron production is clear from Plot's account, and not to be underestimated. Ironmasters and their workers were finely attuned to the processes. Building on practice and opportunities, they were open to new technologies and possibilities, whether those were direct or complementary. Economics was a strong driver here, and one explanation for the growing popularity of air furnaces. Another trigger may have been the increasing quantities of cheap scrap iron available early in the eighteenth century. Large items such as cannon were much more conveniently remelted in an air furnace than in a blast furnace.[31]

After this, as air furnaces spread, other positive consequences were revealed. Ironmasters noted improved quality, with fewer slag inclusions, and enjoyed the

---

30   Schubert, *History of the British Iron and Steel Industry*, 157; Hayman, *Ironmaking*, 19, 86. Hayman is excellent (and accessible) on the detail of iron-making.
31   Plot, *Natural History of Staffordshire*, 160–4, esp. 164 par. 26, describing a ladling process. John Suter offered this reference, and also insights into the possibilities of air furnaces.

flexibility of mixing scrap to produce specific types of iron. In practice, almost inevitably ironfounders would notice that metal left too long in an air furnace partially converted to puddled iron. Air furnaces had other practical advantages for the industry: they decoupled the production of iron goods from the blast furnace, and at a fraction of the cost of a blast furnace. In this they paved a route into ironmaking for people with limited capital, and enabled small iron items such as machine castings to be produced in batches. The classic example of a small-scale start transforming into a giant enterprise is that of Samuel and Aaron Walker, who in 1741 built an air furnace in their backyard. It grew year on year through reinvested profits into the mighty Walker Brothers' works of Rotherham. This is not another post-hoc Victorian rags-to-riches myth, for the partners' annual accounts stand as evidence.[32]

The innovations of Abraham Darby I and Henry Cort were landmarks, and yet the full impact of this ingenuity took some time to materialize, especially in Darby's case. The air furnace example shows that innovation involved apparatus and techniques, and also access to materials and fuel, some notion of economics, and the needs and skills of ironworking communities. There was no simple route to success. So we have to consider what else was happening, and what that explains about the trade and its propensity for technological change. Charcoal or coal, Newcomen or Boulton & Watt steam engines, over-reliance on Baltic bar-iron? All were significant matters for ironmasters. But their decisions were also formed on a human level, assessed according to local and individual circumstances.

Abraham Darby
A sizeable market in cast iron developed during the latter half of the eighteenth century, when it became 'the ideal pre-fabricated material' for bridges and other structural purposes, for armaments, and for a fast-growing new sector: steam-engine components. Cast iron was not, though, considered an industry in its own right, a separate trade from malleable (wrought) iron, before major technological advances were achieved after 1800.[33]

This had been long in coming. A century earlier, Abraham Darby (1678–1717) had made substantial strides in cast-iron manufacture. He is known for systematic trials of coke as a fuel in iron-smelting. Less recognized are Darby's advances in foundry techniques, the more remarkable for being accomplished within a few years, and with little prior experience in iron. Darby started out in malt-milling, before joining three fellow Quakers as partner in a Bristol brassworks. From 1706 he planned a move to Coalbrookdale, Shropshire, chosen for the ready local supplies of pig iron suitable for his new business making

---

32 See John (ed.), *Minutes relating to Messrs. Samuel Walker & Co.*; and below.
33 See Hayman, *Ironmaking*, ch. 7, esp. 92–3.

bellied iron pots. Some of his Bristol brass-founders had applied their skills to the task of casting iron in sand. This, Darby hoped, would replace single-use loam moulds in iron-casting. The method he developed from 1703, casting iron pots in sand with reusable wooden moulds, was sufficiently innovative for a patent to be granted in 1707. He and his heirs thrived on the security of that profitable monopoly, which financed much subsequent experimentation and development of the family business. Darby knew about reverberatory furnaces from his experiences in Bristol, and was among the earliest to apply that technology to ironworking. He acquired the lease of a semi-derelict blast furnace at Coalbrookdale in 1708, and appears to have smelted with coke from the start.[34]

Others had tried this. Thomas Proctor was perhaps the originator, in 1589, though the single-minded Dud Dudley (c. 1598–1684) is better known, having written an account of his trials and discoveries. As an illegitimate son of Lord Dudley, his career was unusual. Oxford-educated, he took charge of his father's ironworks, where from the 1620s he smelted iron ore with coke. The venture was never commercially successful, whatever Dudley claimed later. His product was probably contaminated with sulphur to a point where at red heat it became too brittle to forge.[35]

The suggestion that Dudley was related to Abraham Darby, as a great-uncle of Darby's father, is credible but unsubstantiated.[36] Both were born near Dudley, west of Birmingham, which in Dudley's time was a centre of small ironwares. Later, as Birmingham came to specialize in higher-status goods, its coal and metal-based industries shifted west, and became heavily concentrated in the region known as the Black Country.[37] Whatever Darby and Dudley's relationship, they did not engage directly: the older man died in his late eighties, when Darby was six. But it is possible, even likely, that Darby learned about Dud Dudley's attempts at coal-smelting. Local connections could have led Darby to the part-ruined furnace at Coalbrookdale, which was on the road to Shrewsbury about twenty miles north-west of where he had grown up. The previous operator had absconded following an explosion. Darby took a sub-lease and rebuilt the premises. Had Darby enjoyed Dudley's longevity,

---

34   Hayman, *Ironmaking*, 34–7; (for Coalbrookdale's earlier history and physical development) 133–9; N. Cox, 'Abraham Darby (1678–1717), iron founder, copper smelter, and brass manufacturer', DNB; Raistrick, *Dynasty of Iron Founders*. See also Evans and Rydén (ed.), *Industrial Revolution*, 18–20.

35   P.W. King, 'Dud Dudley', DNB; Hayman, *Ironmaking*, 35; Clow and Clow, *Chemical Revolution*, 331; Dudley, *Metallum Martis*. Ashton, *Iron and Steel*, 10–12, was less impressed by Dudley than was Smiles, *Industrial Biography*, 46–59.

36   It is suggested that Jane Dudley, Dud's sister, was grandmother of Ann Baylies, the mother of Abraham Darby: http://www.pennyghael.org.uk/Darby.pdf. The source has not been verified.

37   Evans and Rydén, *Baltic Iron*, 123.

solutions which eluded the wider iron industry for years to come may have been discovered sooner. He was too early, for instance, to benefit from steam power. He died in 1717, after a two-year illness had obliged him to step back from managing the Coalbrookdale works.[38]

Darby's first years in Shropshire focused on finding the best available coal for coking. The sulphur content of coal in its natural state contaminates iron, so it is unsuitable for smelting. Trials with many different coals confirmed that the most useful for Darby's purposes was the local Coalbrookdale clod, helpfully located close to good ironstone. Local limestone later emerged as a perfect fluxing agent for coke-fuelled smelting.[39] As Darby made progress, fuel became less of an issue, but still the quality variations of raw materials made it difficult to maintain consistency in coke-produced iron. Nor did his process work in forging, converting pig iron into the wrought iron most in demand by the iron trade. But it did the job for Coalbrookdale, where pig iron was used directly to cast household utensils. Darby's make of pots was popular, weighing perhaps a third less than the rival products. By sand-moulding, the pots were further improved. An insertion technique quickly developed, pressing pre-formed legs into the sand to fix them more securely to the pot. He must also have used multi-part moulding boxes. Then, the coke-firing issues having been largely solved after 1713, there followed a considerable increase in production. Darby's final advance was to build a second blast furnace in 1715. His was the only ironworks in Britain organized in this way, with reverberatory furnaces (which separated iron and fuel into different chambers, so that coal could be used without contaminating the product) remelting the iron in order to improve its quality. His system was ergonomic, designed for flexibility.[40]

Darby's was a rich legacy. Cossons hails his 'perfection' of the coke-smelting of iron ore as vying (with Newcomen's engine) to be considered the single greatest technological achievement in ushering in the modern world. The full possibilities of Darby's smelting method, though, were not achieved until more than a generation after his death.[41] This has sometimes been attributed to the disparate, strongly regionalized nature of eighteenth-century British ironmaking. Coalbrookdale was situated in the most innovative of the nation's

---

38   Cox, 'Abraham Darby', DNB; Baggs, Cox, et al., *History of the County of Shropshire. XI. Telford*, 40–56. The world's first operational steam engine was installed near Dudley Castle by Thomas Newcomen in 1712, to drain the coal mines of a later Lord Dudley: Rolt, *The Mechanicals*, 4–5, and facing image. For Coalbrookdale's development as an industrial system, see below, pp. 45–6, 166, 186–7.
39   Cossons, *Industrial Archaeology*, 32, 114; Evans and Rydén, *Baltic Iron*, 251–2; Clow and Clow, *Chemical Revolution*, 332.
40   Hayman, *Ironmaking*, 35–6; Ashton, *Iron and Steel*, 27–34; Cox, 'Abraham Darby', DNB; Cossons, *Industrial Archaeology*, 115; Trinder, 'Abraham Darby [II] (1711–63), ironmaster', DNB.
41   Cossons, *Industrial Archaeology*, 60, 113–15.

Plate 3. Abraham Darby's patented iron pot was the foundation of his company's success. This example weighs about 4kg and has a capacity of almost two gallons. Source: Ironbridge Gorge Museum Trust, HEW 1714.

iron regions, the western Midlands and Welsh borders along the spine of the River Severn. But Darby traded, unusually, in finished domestic goods, the majority sold within his region or shipped out through Bristol to the New World.[42] His business stood somewhat apart from British iron's mainstream.[43] Although Darby was early to spot the possibilities in cast iron, during his own lifetime this branch remained a minority, perhaps even a niche, part of Britain's iron trade.[44] There is another, more convincing, explanation of why Darby's smelting technique was not more widely picked up. Hyde suggests that it worked for Darby because of the economics of his pot, and because he held the patent. The added cost of coke-smelting, compared with using charcoal, was covered by what Darby saved on materials for the unusually

---

42 Hayman, *Ironmaking*, 33; Evans and Rydén, *Baltic Iron*, 252.
43 Trinder, 'Abraham Darby [II]', DNB.
44 Hayman, *Ironmaking*, 36. Benjamin Huntsman undoubtedly drew upon Darby when developing 'cast-steel': see below, p. 50.

thin-walled castings, and by the profitable monopoly on this specific, very popular product. His was a unique situation.[45]

After Darby
It was in the nature of iron production that much innovation resulted from creativity in technique and application. This was not purely about 'hard' technologies. It is also true that solutions which now appear less than monumental could nonetheless fix someone's immediate problem effectively, if not permanently. Darby and Cort were beacons, but they were not alone in their ingenuity. They sit within a long continuum of improvisation.

Iron production, as a strategic industry in a century riven by war, was susceptible to events outside its own control. Tariffs on imports, and the interruption of supplies from overseas, were matters for concern, particularly with demand for bar-iron high and rising. Even so, Baltic iron met the upsurges in British demand for more than a century, its sales there peaking only in 1793. After this, during the wars with revolutionary France, tariffs on bar-iron imports escalated, from £2.81 a ton in 1795 to £5.05 in 1805 and £6.49 in 1813. This was the death knell for both Russian and Swedish iron imports. It was possible only because puddling enabled the malleable iron sector to convert to mineral fuels, setting Britain on a road to self-sufficiency in bar-iron. Effectively the tariffs made Baltic iron more expensive, giving British coal technology a clear advantage during the transition to puddling, which had been comprehensively adopted by 1815. The trade in Öregrund iron was the only survivor of the Napoleonic wars, this prized material still imported for specialist steel-making in Sheffield, Newcastle, and Birmingham.[46]

What, though, of the many decades separating the technological milestones of Darby and Cort? Clearly, ironmaking did not stagnate during that period. Britain's last charcoal-fired blast furnaces were built in the 1750s, and continued to produce significant quantities of iron.[47] How were they replaced, and how was capacity increased? Initiative did not die, as Angerstein shows powerfully. Several well-documented companies, one of them Darby's own, question the view of Ashton and others that British iron production essentially lost impetus in the first half of the eighteenth century. Rydén puts forward a strong case that the way in which this industry was organized gave it tremendous resilience.[48] That he emphasizes organization and not technology is significant.

One ownership model, the Spencer combines formed in northern England to finance and manage clusters of ironworks, will be discussed in more detail.

---

45 Hyde, *Technological Change*, 40–1.
46 Hyde, *Technological Change*, 105–6; Evans, Jackson, and Rydén, 'Baltic Iron and the British Iron Industry', 662; Ashworth, *Industrial Revolution*, 171–2.
47 See above, p. 35.
48 Rydén, *Production and Work*, 15; Ashton, *Iron and Steel*, 235–8 and *passim*.

The Darby company, in contrast, seemingly a solid and conventional family concern, in fact survived many tribulations over its long history. Several times it was revived by its directors' technological leadership, under circumstances which might have left others rudderless.

The Darby company's problem was that its talented principals died young, its salvation the well-resourced and trusted connections who stepped in to protect it. In its darkest hours, after the loss of each of three Abraham Darbys, the business was rescued and protected for a young heir to later inherit. Help came through commercial men, fellow members of the Society of Friends (Quakers), mainly relatives through marriage. If not themselves specialists in ironworking, they found clerks and managers who adapted to the trade and delivered stability.

The period around the first Darby's illness and death brought the greatest danger, with the works then quite new. Charge was first handed to a relative, Thomas Baylies, an interlude which came close to disaster.[49] The business carried heavy debt from the coke-smelting experiments of 1713. Darby had borrowed £1,700 from his Quaker friend Thomas Goldney II (1664–1731), a Bristol merchant, on security of a half-share in the business. Darby had handed over 8 of his 13 shares, and 3 were owned by other Bristol associates.[50] When Darby died intestate, his elder son, Abraham II, was just turning six; his widow Mary died within months. This was a catastrophe in the making.

But Goldney saw the value of a going concern, and, says Ashton, was well disposed towards the Darbys. He introduced his son, Thomas III, recently trained in the Goldney merchant partnership, to look after Coalbrookdale's accounts and sales and production records. Thomas III was granted two shares of his own in 1718, perhaps acquired from the other parties, and his own association with the company continued for fifty years, alongside his many other banking and commercial interests. On occasion he also provided business finance. The other partner, Richard Ford (1689–1745), clerk at Coalbrookdale under the first Abraham, guarded the Darby interest as husband of Abraham II's sister Mary.[51] Of these two young managing partners, Ford took the lead. He knew the business and the industry, and acted boldly to secure the company's fortunes. Sales of pots and kettles doubled within two years. He built a new

---

49 Cox, 'Abraham Darby', DNB.
50 Raistrick, *Dynasty of Iron Founders*, ch. 4, describes these and subsequent business arrangements. Certain of Thomas Goldney II's earlier dealings as a Bristol trader were dubious: Stembridge, *Goldney Family*, 13–17.
51 Cox, 'Abraham Darby', DNB; Trinder, 'Abraham Darby [II]', DNB; Ashton, *Iron and Steel*, 28, 32, 41; http://www.pennyghael.org.uk/Darby.pdf; Wiltshire and Swindon Archives, CRO 473/156, for the Darby and Goldney agreement, 1713. An ironworks clerk was a person of significance, potentially becoming a manager or even partner: see p. 154. For Goldney's role in the slave trade, see p. 165 and fn.

trade in pig iron, and identified growing markets in colliery pumping-engine components. The company began to supply cast-iron cylinders in 1718, first to Thomas Newcomen himself, after which engine parts became core business. Ford also recognized the rising popularity of cast-iron railings, a staple Coalbrookdale product over the following century.[52] Here was resilience on display, the combination of commercial and technical specialism – and access to funds – essential to any successful ironworking organization. Darby's business, thanks to well-disposed and able Quaker contacts, was ultimately very fortunate in surviving, and then thriving.

By another happy chance, Abraham Darby II emerged as an admirable successor to his father. His career illustrates the possibilities to innovate in a mid-eighteenth-century ironworks. He had great vision. Following a Quaker schooling in Lancashire, Abraham II joined the firm in 1728, becoming a full partner ten years later. He demonstrated in 1743 that water running off a waterwheel could be recycled using a steam pumping engine. By ensuring a steady supply of water, this made it safe to operate furnaces throughout the year.[53] This system, first trialled at Coalbrookdale, was adopted for all new coke-fired blast furnaces from c. 1750. In the middle of that decade, still partnered with Thomas Goldney III (and the two strongly linked to John Wilkinson's Bersham ironworks near Wrexham), Abraham II blew in four new coke-fired blast furnaces on satellite sites near Coalbrookdale. Here was a new model in integrated ironworking: having already developed his own collieries, Darby built waggon-ways to transport coal to his furnaces – 'roads made and laid with sleepers and rails as they have them in the north of England', as his wife described. On a moral principle, the second Abraham Darby did not patent innovations.[54]

In combining two industries, coal-mining and ironmaking, Abraham II set an example mirrored by colliery owners, notably in east Shropshire, alerted to the advantages of building their own blast furnaces. Charcoal-fired furnaces were still being built, even as coke-fuelled furnaces announced a new era in ironmaking. In two decades from 1750, 27 coke furnaces started up and 25 charcoal ones ceased operation. Still, there was considerable overlap in old and new technologies. Coke-smelting, which started to appear from c. 1755, was at first the preserve of a few ironworks: Darby built 6 more coke furnaces; Wilkinson added 3; and there were 4 around Merthyr Tydfil. The Carron

---

52 Ashton, *Iron and Steel*, 41–3; Raistrick, *Dynasty of Iron Founders*, 55–7; 129.
53 See Lewis, *Philosophical Commerce of Arts*, section V, 'Of the blowing of air into furnaces by a fall of water'; Newton, 'Farnley Smithies', 160.
54 Trinder, 'Abraham Darby [II]', DNB; Hayman, *Ironmaking*, 39–40; Ashton, *Iron and Steel*, 37, 42, and App. E, 249–52, Abiah Darby's account of her husband's work. It is said that he also declined a Fellowship of the Royal Society: Clow and Clow, *Chemical Revolution*, 331.

ironworks, built on Scotland's central coalfield at Falkirk in 1759–60 and drawing on Abraham II's organizational template, started with five. Its partners recruited the chemical expertise of John Roebuck, who was made a director, and brought in John Smeaton as consultant. Smeaton developed water-powered cast-iron blowing cylinders to provide blast for Carron in 1760, attempting to solve the inadequacies of blowing bellows.[55] In 1754, when Charles Wood and Gabriel Griffiths visited eleven west Midlands forges, Darby's had been the sole example noted as using coke pigs. Very soon afterwards came a striking upturn in their use in forging, as masters weighed the falling price of the coke product against the increased work required to turn it to bar-iron, and saw the scales shifting towards coke.[56]

The Carron Company was a source of important advances in technique and technology, from both managers and shop-floor workers. In bringing together smelting, founding, and forging on one site, it had further honed the idea of a modern ironmaking plant. With technology advancing there and elsewhere, as Hayman describes, by the 1780s a new type of ironworks had been created, with its own sources of iron ore and fuel, a structure that came to dominate ironmaking in the nineteenth century.[57]

Abraham Darby II had left his own technical and organizational legacy. The most commercially successful of the three eighteenth-century Abraham Darbys, he was also the longest lived, though he died at fifty-one. Certain of his achievements have been disputed, either attributed to his father or downplayed. Beyond any doubt, though, his was a remarkable career, his breakthroughs equipping a following generation to push through the next obstacles.

The experience of mid-eighteenth-century ironmaking – in its highs and lows, its developments and its frustrations – was far from one of inertia and stagnation. In fact, this phase reveals very much about the nature of innovation, as it happened in real time. Was there resistance to change? There was certainly conservatism in the industry, for ironmasters were after all heavily invested in existing technology and practices.[58] But also very evident was an air of urgency in certain quarters, about addressing external and political concerns just as much as cracking the challenges of technology. Regionalized the industry might have been, functioning somewhat differently in different localities, yet a healthy

---

55 Hyde, *Technological Change*, 54–6; Hayman, *Ironmaking*, 39–40; Ashton, *Iron and Steel*, 36–8, 48–51; Ashworth, *Industrial Revolution*, 170. Smeaton was only partially successful, for it took steam-powered blowing to make coke blast furnaces efficient: Cossons, *Industrial Archaeology*, 115. For Roebuck and Smeaton, see ch. 3.
56 Hyde, *Technological Change*, 54–6; Hyde, 'The Iron Industry of the West Midlands', 39–40; Ashworth, *Industrial Revolution*, 170.
57 Hayman, *Ironmaking*, 40–5; Norris, 'Struggle for Carron'.
58 See, for instance, Hayman, *Ironmaking*, 60.

transmission of ideas, techniques, and organization is clear, and with that came the adaptations and understanding to deliver further progression.

Henry Cort

Angerstein noted several trials to improve forging techniques, including those by Charles Wood in Egremont, Cumberland, whom he visited in 1753. Wood and his brother John, of Wednesbury, Staffordshire, also an ironmaster, developed a significant method of remelting coke pig iron in a foundry furnace to remove the silicon. Known as stamping and potting, it was patented in 1761 and 1763. Stamping and potting was key to increasing output in the forge sector – accounting for almost half of bar-iron production in 1788 – before puddling and rolling became viable in the 1790s. It was widely used across the iron districts, persisting into the nineteenth century even after puddling had become mainstream. Charles Wood afterwards played a role in establishing Cyfarthfa ironworks in Merthyr Tydfil, when its promoters Anthony Bacon and Wood's brother-in-law William Brownrigg, both from Whitehaven, Cumberland, commissioned him to install a reverberatory furnace and stamping and potting forge there in 1766. Wood recruited a small core of furnace- and forgemen from Shropshire and Cumberland, highly skilled across the range of ironworking. He subsequently managed Cyfarthfa works until his death in 1774.[59]

Patents resulted from other trials where mineral fuel was used to make bar-iron. But patenting was cumbersome, and sometimes an expensive mistake. An evidently promising venture was developed in 1766 at the Coalbrookdale company by two Quaker employees, Thomas Cranage, forge supervisor, and his foundryman brother George, backed by the company's interim managing partner, Richard Reynolds.[60] The Cranage method involved heating iron and fuel separately in an air furnace, a system then increasingly applied in forges, and used a forging technique similar to puddling.[61]

Reynolds had been enthusiastic. It was 'a matter of very great consequence', and after the first trial 'the success surpassed the most sanguine expectations'. The iron drawn out 'is the toughest I ever saw', he wrote to Thomas

---

59 Evans, *Labyrinth of Flames*, 57–9, 16, 205; Hayman, 'Shropshire wrought-iron industry', 55–6, 62–4, 108–13; *Ironmaking*, 42–5; Hyde, *Technological Change*, 77, 83–8, 92–4. The Wood patents were BP 759 (1761) and BP 794 (1763). For Wood, see pp. 80–1.
60 Raistrick, *Dynasty of Iron Founders*, 86–7; also 131–6; *Quakers in Science and Industry*, 135–6. For Reynolds: Milligan, *Dictionary of British Quakers*, 357–9; Trinder, 'Richard Reynolds, 1735–1816' and 'William Reynolds (1758–1803)', DNB; Bruton, 'Shropshire Enlightenment', 106, 113 and *passim*. Cranage Forge, Cheshire, built in the 1660s, was an important ironworking centre, supplier of rods to Chowbent, and owned at one time by Denis Hayford, though no connection of these Cranage brothers to the forge is evident: see Awty, 'Charcoal Ironmasters'.
61 Ashton, *Iron and Steel*, 87–90; Hayman, *Ironmaking*, 43, 47; Bruton, 'Shropshire Enlightenment', 123; Smiles, *Industrial Biography*, 86–8; BP 851 (1766).

Goldney. Yet when William Lewis visited the Ketley ironworks in 1768, he found Reynolds regretful. The patent had been

> obtained too rashly, before they had considered the expence, arising from the great waste of iron, one half being lost in the operation, whereas with charcoal the loss is only about one third.

What is more, the extreme heat rapidly caused failure of the furnace bottom, even when iron-plated. The Cranage scheme turned out to be impractical, on grounds of both cost and quality.[62]

The Cranages have possibly been overrated as innovators. There was, thinks Hayman, undue deference by Raistrick and other writers towards the legacy of the Darbys and their close Quaker associates, including several generations of Cranage. Thomas Cranage had been at Carron ironworks in its early days, around 1760, and was highly experienced. But innovation within the iron trade was often not 'invention', if that is shorthand for the product of a single mind. A step forward like the Cranage technique was formed by long periods of trial and discussion among workmen at Coalbrookdale and elsewhere. Here, the detail is hard to pin down, the process described only vaguely in the patent. The episode does not, though, reflect well on Richard Reynolds' technical judgement.[63]

Henry Cort came from an unlikely direction, as a naval agent in Hampshire. Made manager of a family-owned forge which supplied iron goods to the navy, he looked to expand further into that lucrative market. But the Navy Board continued to insist on using Swedish bar-iron. Cort understood that the policy would change only if a domestic product became available that matched or bettered Swedish quality. He saw, too, that political pressure would require the Board to buy British, all else being equal. The potential profit for whoever solved this was enormous.[64]

After lengthy trial and investigation, Cort registered patents in 1783 and 1784. His process brought together new techniques and manual skills, combined with existing technologies. Pig iron was refined in a reverberatory furnace, with scrap added. These furnaces transmitted heat from above, so that a current of air blew down onto the iron. As the iron rose to white heat, a blue flame indicated that its carbon content had started to react with oxygen in the air. The judgement and skill of the forgemen, who stirred the molten metal to maximize this chemical

---

62   Raistrick, *Quakers in Science and Industry*, 136; Gibbs, 'Notebook of William Lewis', 208.
63   Hayman, 'Shropshire wrought-iron industry', 12–13, 55–61; also 'Cranage Brothers'; see p. 72.
64   C. Evans, 'Henry Cort (1741?–1800)', DNB; Hayman, *Ironmaking*, 46–9; Evans and Rydén, *Baltic Iron*, 270.

process and then removed it as a spongy mass from the furnace, were critical. It was afterwards shingled under a hammer, and reheated to pass through a rolling mill, eradicating any remaining impurities. Cort's achievement was in assembling many existing practices into an effective system.[65]

While Cort energetically promoted his method, it was not at first well received. Ironmasters and workers were sceptical because Cort lacked pedigree in their trade. Several were angered that their own technologies, some patented, had apparently been borrowed without consent. In certain instances, that was a fair comment. It does, though, seem that Cort largely arrived at the puddling process through his own efforts. Reportedly, only several months after registering his rights in 1784 did he first hear of the Cranages' earlier patent.[66]

In 1789 Cort's cause received a great boost. The Navy Board announced that all potential suppliers of bar-iron must manufacture according to his patented system. But very quickly, dubious dealings came to light. Up to £50,000 from official balances had been channelled via Adam Jellicoe, deputy paymaster of seamen's wages, into Cort's business, in which Jellicoe's son Samuel had been made a partner. Furthermore, the elder Jellicoe had himself convinced the Navy Board to insist that contractors use the Cort process. All the while, Jellicoe held the patent as security on his, or more accurately the government's, loan to Cort. This arrangement may have been known to the Navy Board directors, but the blame fell lower. Whether Cort was complicit, or even aware of it, he was ruined, held partly responsible for the missing funds, and bankrupted to the tune of £27,500. Friends saved him from jail, but he remained an undischarged bankrupt until his death. The patents were confiscated, becoming government property, but no effort was made to recoup royalties.[67]

That in brief is the story of Cort. Questions remain about his process and the speed at which it was taken up. It was a great revelation, long in coming, and promising to revolutionize how wrought iron was made. It was in practical terms unpatented from 1789, yet not widely embraced, or not immediately so. And what of the enigma of how iron was being made in Britain from the 1750s: that the last charcoal furnaces were built, many coke-fired ones appeared, and a boom in ironmaking ensued, all before Cort came on the scene?

Seizing Cort's puddling and rolling patents had effectively cancelled them. With the government evidently more interested in punishing him, any income from this increasingly valuable asset went uncollected. But the patent continued to exist, so no one else could claim the rights for themselves. Even had things turned out differently for Cort, the patents could have been challenged, for his method built on processes like that described by Plot in the 1680s – serendipitous,

---

65 Hayman, *Ironmaking*, 46–7; Ashton, *Iron and Steel*, 90–3; Ashworth, *Industrial Revolution*, 170–1.
66 Evans, 'Henry Cort'; Hayman, *Ironmaking*, 47–8; Ashton, *Iron and Steel*, 90.
67 Evans, 'Henry Cort'; Ashton, *Iron and Steel*, 93–4; Percy, *Metallurgy*, 630–1.

but long known. For the same reason, as ironmasters already used elements of the process, enforcing against piracy may have proved impossible.[68] Yet there remained impediments to taking it up, most obviously that the unproven Cort process required industrial trials. To adopt it meant investment, not least in retraining forgemen and persuading supervisors to accept new approaches.[69]

Being patent-free is not sufficient on its own to make a technology attractive. Pioneering, promising processes, patented or otherwise, can take years to reach their potential and find their niche. In this industry, there are prominent examples pre-dating Cort. Neither the second Abraham Darby nor Benjamin Huntsman, both Quakers, used patents. Darby's father took only one patent, for his most marketable concept, the iron pots. As described, the older Darby's smelting method took decades to be absorbed within the trade. Huntsman, in developing the cementation process to produce crucible (cast) steel, built on the first Darby's discovery that coke could be used as a metallurgical fuel. After experimenting from the 1730s, and producing his (and Europe's) first steel ingot *c.* 1742, Huntsman set up in business as a steelmaker in Sheffield in 1751. Tweedale sees Huntsman's achievement as 'truly revolutionary', marking the beginning of the modern steel industry. The steel produced was costly and highly specialized, uniform, and took a cutting edge, so was prized for edge tools. Huntsman spent years identifying the optimum types of furnace, flux, and fuel. So why was his process so slow to catch on? Crucible-steel making made punishing demands on the workforce, in physical strength, experience, and judgement as well as manual skill. The human element was a major stumbling block in bringing the system into wider use. Sheffield cutlers, Smiles suggested, were resistant to working a material which was so much harder than they were used to, so that Huntsman himself at first had difficulty finding a market. Huntsman's business was not a great commercial success: he was inclined to protect his reputation for exceptional quality, which was always challenging. He was also secretive about elements of the process. These factors deterred, or at least delayed, competitors.[70]

Second, why did the balance towards coke shift quite decisively from the 1750s? Coke-smelting predominated by 1775, with about thirty coke furnaces then at work, producing half of the country's pig iron.[71] Riden's estimated number (with approximate total tonnage) of British coke-fired furnaces producing pig iron shows an increase from 4 (2,000 tons) in 1750–4, to 26 (19,000) in 1765–9, 56 (50,000) in 1780–4, and 72 (70,000) in 1785–9. Over that

---

68 Ashton, *Iron and Steel*, 94.
69 Hayman, *Ironmaking*, 48; Hyde, *Technological Change*, 95.
70 D. Hey, 'Benjamin Huntsman, (1704–76)', DNB; Hey, 'South Yorkshire Steel Industry'; Smiles, *Industrial Biography*, 102–11. For more about Huntsman's process and market, see Tweedale, *Steel City*, 28, 32–4, 38–42.
71 Hyde, *Technological Change*, 69–70.

full period, charcoal furnace output of pig iron, which had hovered around the high twenty-thousands since 1720, fell from 26,000 to 10,000 tons. The figures for bar-iron output are more fragmentary, and not clearly comparable over time. A survey of 1788 (though perhaps based on older data from the earlier 1780s) showed 16,400 tons of bar-iron made in a year according to the 'Old Plan', with 15,600 tons coming entirely from coke-fired processes. Riden attaches greater weight to evidence presented to a parliamentary committee in 1785, which assessed England's annual bar-iron output as between 20,000 or 25,000 tons and 30,000 tons, 'and it is a very increasing article'.[72]

What is to be made of this? Several factors were at play. Much of the coke-smelted pig was destined for casting directly into products to feed booming markets in armaments, steam engines, or domestic goods, where the forging problem was not directly relevant. The new coke-fired blast furnaces recorded in Yorkshire after 1750, many of them around Sheffield, and a little later in Leeds, were wholly or mainly turning out cast-iron products. The first was at Walker Brothers' Masbrough works in Rotherham (1765), where charcoal furnaces also continued in use. Afterwards came Chesterfield (1777) and Renishaw (1782), both in north Derbyshire. Six large coke furnaces were built at Swallow's new ironworks in Chapeltown, Sheffield (1778). At Seacroft, Leeds, four coke furnaces operated from 1776.[73]

For bar-iron producers too, economics were shifting in favour of coke-smelting, a move further encouraged by the advent of stamping and potting. This process found its time once coke replaced coal-plus-flux. Further incentive came with the adoption of what Hyde calls 'supportative' innovations. The *Oxford English Dictionary* (OED) notes 'supportative' as a neologism, and a superfluous one which says nothing that 'supporting' or 'supportive' does not already, and more neatly. Yet it has a purpose here, and in following chapters, in identifying a concept and Hyde as the source. The term suggests something more than ancillary, or just helpful – 'supportative' could suggest a facilitator, akin to a catalyst.

Such technologies were often associated with prolonging the lifespan of an ageing system. Darby's pioneering use of steam power to feed his waterwheels, from 1743, became a standard method during the 1750s. This, the application of Newcomen-type technology twenty years or more before Watt's improved steam engine, was pivotal to his industry's expansion. Even after 1775, many Newcomen-style systems remained in operational use where fuel economy was not the primary concern. Where local water power was sufficient, blast furnaces

---

72 Riden, 'Output of the British Iron Industry', 442–59, and esp. Table 2. The sources did not always make clear whether Scotland and Wales were included.
73 Mott, 'Early Ironmaking', 225–35; Sellars, 'Iron and Hardware', 395; John (ed.), *Minutes relating to Messrs. Samuel Walker & Co.*, 6–9; Riden and Owen, *British Blast Furnace Statistics*.

could still work from a single waterwheel, unsupported by steam. Remarkably, Walkers of Masbrough, on the River Don, an industry leader in many respects, was among these: new furnaces were built in 1765 and 1779, yet the site had no steam engine until 1782.[74]

A second significant 'supportive' technology was Smeaton's blowing cylinder, developed for the Carron Company. The effects of Carron as a model ironworks, says Schubert, were felt across the whole trade. The new works were built in 1759–60 with only coke furnaces, and Smeaton was afterwards commissioned to improve the blast. By Schubert's estimate there were then – at most – seventeen coke furnaces in blast in the whole of Britain. This increased by 14 in the early 1770s, and rose to 81 in 1790, when only 25 charcoal blast furnaces remained active. There are discrepancies in the figures, but the trend is clear: a marked rise in coke-smelted ironmaking some time ahead of Cort's appearance on the scene.[75]

How much of this is attributable to Carron's example? It was important, no doubt, and especially in sparking similar ventures across southern Scotland. But change was already evident, and precursors to the Cort method were in use, taking elements of puddling and other new forging techniques. Any lingering prejudice against coke-smelted iron was dissipating by the mid-century. Most importantly, ironmasters were then balancing possibilities based on their own local circumstances, and trying to weigh costs while sustaining quality.[76] Technological diffusion generally carries a time lag, particularly in a highly capitalized activity such as ironmaking. Here, such delays shortened markedly as the eighteenth century proceeded.[77] Choices became simpler once technology settled and proved itself.

A final complication relating to the uptake of Cort's process came in its timing. While prospects of war and of high import tariffs created economic uncertainty, they also gave incentive to push forward with a technology of increasing promise. Take-up was slow before the 1790s. But during that decade, British ironmaking turned rapidly to mineral fuel. The balance tipped in favour of home-produced bar-iron, and Cort enabled it. South Yorkshire ironmasters had lobbied forcefully in 1756 to oppose tariff-free iron imports from the colonies.[78] Their successors faced escalating surcharges on supplies

---

74 Hyde, *Technological Change*, 69–75, 194; Riden and Owen, *British Blast Furnace Statistics*, 129, show the Masbrough works under the names of its furnaces, Holmes, Park Gate and Milton.
75 Hyde, *Technological Change*, 194; Schubert, *History of the British Iron and Steel Industry*, 332–3.
76 Rydén, *Production and Work*, 29–31.
77 Hyde, *Technological Change*, 195–6; also 63–75.
78 Sellars, 'Iron and Hardware', 395.

from the Baltic, and were conscious that war had stiffened government resolve.[79] Ironmasters and iron-buyers had choices to make, in order to secure their businesses for the longer term. Events in the 1790s encouraged investment – in new methods, in the necessary premises, in plant and retraining the workforce. But this was not an inevitability. Cort's technology had to be demonstrated as viable in practical terms, and the effort to make it an economic success would follow.

Measuring choices
Cort's integrated method laid a pathway to make bar-iron using coke-smelting. It was transformative, though achieving it took some years. Richard Crawshay was key, adapting puddling and rolling technology into an unquestionably superior system through his large-scale trials at Cyfarthfa during the late 1780s. Even late in 1791, four years on, Crawshay had contemplated abandoning the mission and resigning himself to stamping and potting rather than puddling.[80]

Crawshay, born in Yorkshire's West Riding, a man of huge ambition, amassed his first fortune as a London merchant trading in Baltic iron. He had part-owned Cyfarthfa ironworks for a decade, though inactively. When the managing partner died, Crawshay saw a timely and profitable opportunity. He took full control in 1786, shortly after Cort's patents were registered. Crawshay's heavy investment transformed Cyfarthfa, a decade later, into by far the largest ironworks in Britain.

Known locally, and not altogether approvingly, as 'the Iron King', Crawshay let nothing stand in his way. His combative approach attracted him to puddling, which he saw as a potential opportunity to replace troublesome forgemen with more tractable workers. Streamlining the works promised further division of labour, with economic benefits as well as greater control of employees.[81]

Only a year into Crawshay's career as ironmaster, he contacted Cort, and agreed to buy into the new process. Cort was promised 10s. a ton on all bar-iron produced at Cyfarthfa using his system. At the time of his downfall, Cort had received nothing, and later died in poverty. Crawshay could not then produce much puddled iron that was usable, but it seems that his agreement with Cort was not honoured. His arguably unethical behaviour – Percy called it 'the shabbiest part of this melancholy story' – does not detract from the fact that in backing the trials, Crawshay made the technology work on a commercial

---

79 Evans, Jackson, and Rydén, 'Baltic Iron and the British Iron Industry', 645.
80 Hyde, *Technological Change*, 96–101. Hyde painstakingly worked through Crawshay letter books and archives, now deposited in Gwent RO, and others in south Wales. Cyfarthfa is pronounced kuh-VARTH-va.
81 Evans and Rydén, *Baltic Iron*, 260–71, offer more detail on Crawshay's trials and other developments mentioned here.

Plate 4. Penry Williams, 'Crawshay's Cyfarthfa Ironworks', c. 1827. The civil engineer Watkin George made significant changes at Cyfarthfa from the mid-1790s. Most striking is his 100-ton waterwheel, Æolus, here providing blast to the five furnaces lined up to the left. In front of each furnace is a casting house where molten iron was tapped. The large multi-chimneyed building on the right, the forge, housed puddling furnaces. In the background, smoke rises where coal burns in open heaps to produce coke. Source: Amgueddfa Cymru – Museum Wales.

scale. His great gift to the trade was in spotting, in this untested method, the potential to transform iron-refining.[82]

Exceptional progress was reported in 1788, though subsequently matters took a downward turn. The blame for failures before the autumn of 1791 was squarely placed by Crawshay upon his junior partner and works manager, James Cockshutt. Cockshutt, a member of the Wortley Forge family in south Yorkshire, had solid credentials in iron manufacture. That included a decade as manager of Pontypool Forge, then the largest iron refinery in Britain. The Cyfarthfa workforce struggled to adapt to Cort's technique, and so did

---

[82] Hyde, *Technological Change*, 95–102; C. Evans, 'Richard Crawshay, 1739–1810', DNB; Percy, *Metallurgy*, 632–9, gives a long account of the episode.

Cockshutt. Crawshay dismissed him, dissolving their partnership in September 1791. Crawshay evidently then went fully hands-on and made critical changes to the trials in 1792–3.[83]

It was fortuitous, and maybe no coincidence, that Crawshay acquired the Cyfarthfa works just as Cort's process cried out for industrial testing. Cort had previously approached the technologically progressive Carron Company. Carron produced mainly castings, though some iron was forged there. The managing partner, Charles Gascoigne, in full charge since 1776 and a keen experimentalist, received Cort in 1786. But Gascoigne had already settled on a system which suited him, using charcoal to produce bar-iron from Carron pigs. Cort was rejected, and perhaps this was for the best, as Gascoigne was very soon afterwards bankrupted. He fled to Russia with a dozen of Carron's skilled workers, and neither he nor the company was then able to help Cort, nor perhaps help themselves.[84]

Cort's experience was not unique. In looking for reasons why good innovations take time to diffuse, the same themes recur: technologies need to be polished, or radically rethought; potential users are already invested in other methods; or they do not have financial or managerial capacity to engage at that point; or they are working on their own solutions. With hindsight, Cort's technique was the obvious choice. At the time it was much less clear. Perceptions, needs, experience all differed; in rejecting it, contemporary ironmasters perhaps reacted quite rationally.

One such was John Wilkinson. Practical, innovative, and then at the height of his career, he is best known as supplier of engine cylinders to Boulton & Watt. In parallel with Cort's and Crawshay's trials, he too was investigating methods to convert coke pig to wrought iron. By 1789, Wilkinson had improved his rolling process and abandoned the second stage of forging (chafery). Puddling experiments, in progress in 1790, took time to bear fruit, with Wilkinson on the brink of giving up. But he made breakthroughs in 1791, and by the middle of that decade had a fully operational system.[85] With Cort's methods then widely known, Wilkinson must have been aware of the cancelled patents, and influenced in some measure by their content.

The Cockshutts of Wortley were early to adopt Cort's process, and were probably the first in Yorkshire to make wrought iron profitably and on a large scale. James Cockshutt (1742–1819) took puddling and rolling there around the time of his sacking from Cyfarthfa in 1791. Succeeding his brother as manager

---

83   Cookson, 'Wortley Forge'; Evans, *Labyrinth of Flames*, 18, 63–6 and *passim*; Hyde, *Technological Change*, 96–101; Hayman, *Ironmaking*, 48–9. See chs 6 and 7, below.
84   M. Chrimes, 'Charles Gascoigne, *c.* 1745–1806', BDCE, 244–6; Robinson, 'Charles Gascoigne, *c.* 1738–1806', DNB. The birthdate of 1738 appears correct.
85   Harris, 'John Wilkinson, 1728–1808', DNB; M. Chrimes, 'John Wilkinson, 1728–1808', BDCE, 783–4; Ashton, *Iron and Steel*, 63–86.

of the Wortley complex, James installed a puddling furnace at the Low Forge, and what was described as 'the first bar mill with grooved rolls' in Yorkshire to roll iron bars by Cort's method.[86] Afterwards, James enlarged the works with puddling furnaces, foundries, machine shops, a tin and sheet mill, and several wire-drawing mills. Much of the forge's production was sold as bundles of rods, supplying a large local market of nail-makers in Ecclesfield and other neighbouring parishes.[87]

So with puddling's rapid adoption during the 1790s, the full impact hit the iron trade after 1800. As forging entered this new era, cast-iron production became recognized as a trade in its own right.

Building resilience

To summarize how British ironmaking developed: already a major sector in 1700, it progressed from heavy reliance on 'forest economies' and Baltic imports to high levels of national self-sufficiency, with production well integrated and an almost complete shift to mineral fuels by the century's end.[88] This was achieved through sustained efforts and many frustrations, which ultimately established viable and satisfactory ways of working

The narrative did not start or end there. Innovation flowed within a multi-process industry, advancing through series of long, sometimes tortuous, explorations. Even puddling was not an end, but the start of a new phase which in turn threw up fresh problems to be worked around, and worked out.

Nor was innovation purely about technology. Far from it, for the accommodations required in reshaping ironmaking took many forms, and intersected with new ways of working in other trades. One industry's product is another's prized material, and once iron became available that was suited to new applications, this had rapid impacts. In particular, new qualities and quantities of iron were fundamental to how machine-making and engine-building progressed, and they presented new structural possibilities for factories and bridges. Entirely novel objects and systems – the powered machines and multi-storey factories which spring to everyone's mind at the mention of 'industrial revolution' – grew on the back of iron's eighteenth-century achievement.

Yet this consideration, 'what happened next?', is only one perspective on iron's own revolution. The path taken by this or any industry was not pre-ordained. The experiences of continental ironmaking substantiate the point. Britain was

---

86   Andrews, *Story of Wortley Ironworks*, esp. 47.
87   Cookson, 'Wortley Forge'; Hunter, *Hallamshire: A New Edition*, 170 fn.; Ashton, *Iron and Steel*, 96–7; Hayman, *Ironmaking*, 48; Andrews, *Story of Wortley Ironworks*, 5, 17, 44–50.
88   The phrase is from Evans and Rydén (ed.), *Industrial Revolution*, ch. 1.

a rarity in having coke-smelting before puddling. Because puddling and rolling were not tied to a fuel type, they could drop into a traditional forge framework where the iron was made by charcoal. In places, puddling actually extended the life of charcoal technology. Not until 1853 did coke-smelted pig iron overtake the charcoal product in France, while Italy had no coke-smelting at all until 1899. This was all about fitting local circumstances. After all, Britain itself had a long track record in contingencies, its own 'coal technology revolution' having started falteringly in the sixteenth century. Dissemination had always been uneven.[89]

The age threw up its own paradoxes. One of the great compromises was stamping and potting, the imperfect provisional arrangement. Indeed, puddling and rolling did not directly descend from this technique. Rather, it drew from it, and from a variety of other earlier, not fully satisfactory, discoveries. But stamping and potting provided a temporary scaffolding, enabling other aspects of the iron trade to develop around this system which was simultaneously 'best possible' and also a constraint. Consequently, a supporting and expandable infrastructure had been created, into which puddling could quickly settle.

Nor were there fixed points of technological excellence. Influential ideas sometimes came out of left field, from unexpected directions – Cort was a prime example – while the companies best known for being progressive were not consistently so. Life, or more likely in the eighteenth century, death, intruded on company succession and the potential to innovate. In this the Darby company was particularly ill-starred, though also fortunate in surviving and reviving, and more than once. Similarly at Wortley, consistently resourceful through the Cockshutt era, a rapid decline and break-up of the business followed James's death in 1819.[90]

Initiative was not an outcome only of maverick ingenuity (Cort) or leading companies' consistent determination (Darby, Cockshutt). Some long while before the 1790s' dramatic breakthroughs, technological advances in ironmaking already supported the trade's drive for more flexibility and responsiveness. This was the transformation Angerstein witnessed in progress during the early 1750s, and it did not end there. This explains the paradox of how there could simultaneously be a boom in British iron production after 1750 and a supposed technological stalling before puddling came on the horizon in 1783. The answer is that there was no stalling. Innovation was unceasing. Failing to find definitive solutions in fact gave a greater incentive to continue striving. Even in 1790, who could know whether Crawshay's solution would surpass Wilkinson's, or if one of the other trials in progress would outperform them both? And that principle holds true for all technological initiatives in the previous half-century.

89   Evans and Rydén (ed.), *Industrial Revolution*, 8-9, 15–16.
90   Cookson, 'Wortley Forge'.

The case of the cupola furnace is a real-time example of what was happening on the ground, an influential yet understated technology which opened opportunities in ironmaking from around 1750. Later, the cupola acted as a 'supportative' technology for machine- and engine-building. It was not a new concept in 1750, but it was versatile and found its time. With charcoal-smelting on the wane and iron-founding literally emerging from the woods, coke-fired blast furnaces opened new geographical horizons. These options expanded further with steam power, or steam-assisted waterwheels. In consequence, furnaces moved closer to sources of coal and iron, and also to the main markets, whether those were for cast-iron goods or bar-iron.[91] The cupola was a next stage, offering more flexibility in location, organizational structure, and finance.

The cupola was a small blast furnace, used not to smelt but to remelt. With this technology, a foundry of modest size could cater for local specialisms, and at a fraction of the capital cost of a full-scale ironworks. To machine-makers, this presented particular advantages. Castings from reverberatory furnaces were hard but brittle. In contrast, the cupola's output of cast iron was fluid, less likely to fracture, and, according to Ashton, 'peculiarly suited to the production of machine parts'.[92] Later, leading machine-makers and engine-builders built foundries within their own integrated machine-making operation. But before that, cupola furnaces were installed in local foundries, and offered an invaluable facility – one which would long continue – to small engineering businesses who needed castings, often custom-made, without making a large investment.

Because cupola technology was transportable to any location within reach of coal supplies, it was brought into use away from the traditional centres of iron production. The impact on mechanical engineering was rapid: machine and engine parts could then be manufactured in northern towns, on sites close to machine-makers and their textile-manufacturer customers. This clustering – of textiles, machine-making, and iron-processing – came to be associated with dynamic innovation in northern textile centres. This is no small consideration. The economies of having suppliers, subcontractors, and skilled workers near at hand were significant, enhancing a business's ability to respond. Equally important was the physical closeness to customers, especially so in the early days of trial and error, when flexibility and alertness to the needs of machine-users were imperative.[93]

---

91   Hyde, *Technological Change*, 121–3.
92   Raistrick, *Industrial Archaeology*, 46–8; Hayman, *Ironmaking*, 86–9; Ashton, *Iron and Steel*, 102–3. The cupola attributed to William Wilkinson, who patented a cupola furnace in 1793, was probably designed by his estranged half-brother, John. Versions of the cupola were, though, around much earlier, and the Carron ironworks considered installing one from the start, 1759–60: Chrimes, 'Charles Gascoigne', BDCE. See also Harris, 'William Wilkinson', DNB.
93   Cookson, *Age of Machinery*, esp. 198–202.

In Leeds, the first such compact stand-alone foundry appeared before textile-engineering had emerged as an industry in its own right. The new venture, called the Cupola Company when launched in 1770, first served the Middleton Colliery and its waggon-way, and other south Leeds collieries. The foundry's own waggon-way sidings also allowed easy access to the town's improved waterways and the booming urban centre, and it soon developed a wider customer base. This quickly became a significant part of Leeds's industrial infrastructure, exemplifying a wave of new interest in the west Yorkshire towns. Its owners were themselves in-migrants as well as investors, bringing vital skills from Wortley and Sheffield, and attracting other migrant workers with skills that were foundational to the town becoming an engineering centre.[94]

That regional boom – urban and mechanical – had the further effect of breathing new life into a set of much older furnaces and forges. The Wortley-centred Spencer syndicates, introduced in chapter 1 and to be discussed in detail in chapter 6, were fragmenting. The last Spencer withdrew from all iron-trade interests in 1765, by which time Wortley itself was prospering under the Cockshutts.[95] On the fringes of Leeds were two semi-moribund ironworks, formerly outlying sites in the Spencer combines. They too revived under new and more progressive ownership, finding a new management in tune with local demand. Seacroft Foundry drifted from view by 1742 but reappeared later as a discontinued 'coak furnace', so perhaps the scene of coke trials. Seacroft was rebuilt in 1780, presumably by Eyres, Wigglesworth & Co., who commissioned a Newcomen-style pumping engine from John Smeaton in 1781. In 1790 it had four coke furnaces in operation.[96]

The other Leeds site, Kirkstall Forge, part of the Spencer empire from 1676, underwent a renaissance from 1779 under Butler & Beecroft, assisted by the new Leeds–Liverpool waterway passing nearby. A rapid programme of building and refurbishing plant and waterwheels was followed in 1797 by the installation of puddling furnaces from Samuel Witton of Sheffield. The partners had already expanded into finished goods, where the profits were high, alongside the forge's core business of rods, bars, and plates. With housing and workshops for skilled workers added, Kirkstall took on the character of an industrial community.[97]

An affinity suggests itself about this new breed of ironmasters: receptive to the market – in fact, positively seeking fresh opportunities – while also open to process innovation. Is this comparable to the changes then occurring in Leeds

---

94 Cookson, 'Hunslet Foundry'.
95 Cookson, 'Wortley Forge'; Raistrick and Allen, 'South Yorkshire Ironmasters', 177; Butler, *History of Kirkstall Forge*, 7; Awty, 'Spencer Family, per. c. 1647–1765', DNB.
96 Raistrick and Allen, 'South Yorkshire Ironmasters', 169; Riden, *Gazetteer of Charcoal-fired Blast Furnaces*, 105; *Bailey's Northern Dir.* (1781), 222; WYAS Bradford, SpSt/5/5/2/15; Scrivenor, *History of the Iron Trade*, 359; 'John Smeaton', BDCE, 628.
97 Butler, *History of Kirkstall Forge*, 7–8, 16–17, 46–7; and see pp. 190–1.

textile merchanting? It was not a complete overturning of an old guard, but a sufficient infusion of new blood and ideas to introduce new perspectives. In neither trade were the pivotal agents of change complete newcomers. Some were former junior partners (Cockshutt, Gott) or merchants with resources and substantial knowledge of commerce and specific markets (Crawshay, Elam, Marshall, the Kirkstall partners). The pattern was not limited only to Leeds and its neighbourhood.

The revivals and launches of ironworks in the 1770s were followed by a new generation of coke-fired ironworks. In west Yorkshire a significant cluster formed on the coal and iron measures south and west of Bradford, linking by waggon-way to the Bradford Canal: the mighty Low Moor Co. (1791) joined Birkenshaw (established by 1782), Bowling (1788), Shelf (1794), and Bierley (1811).[98] Decisions behind these ventures, the large investments in very new *process* technologies, reflect the levels of manufacturing activity already in the region. The ironworks themselves then facilitated further kinds of industrial enterprise.

Iron *products* were themselves advancing technologies, proving their worth in steam engines and factory-building, in transport, and then in textile machinery and machine tools. The iron trade, in transforming itself, was also transformative across all branches of engineering. Iron supported the most innovative ventures: steam-driven machinery and iron-framed factories, bridge-building and power systems. The most spectacular icons of advancement are associated with cast iron. But the forge sector was no less influential. Malleable iron fed a booming market in the highly specialist area of fine working parts for machinery, notably textile and machine-tool engineering, and much else too. The influence of the iron revolution, long in coming, was felt in every part of life.

There is still much to discover about the eighteenth-century iron trade, in its various manifestations. Even about some celebrated individuals and companies, little is known. There is more understanding of iron's general context. In transforming itself, it built capacity, and also resilience – and notably adapted to local conditions and changing circumstances. While there were laggards, British ironmaking overall responded well during a century of tribulation, when external events added to the usual domestic concerns. The charcoal issue was a major stress, and became part of an impetus for organizational as well as technological change. Throughout, innovation was a reciprocal process, connecting with, and learning from, other iron producers and other industries, and beyond national boundaries. It was pragmatic and grounded in existing practice, setting a path adapted by other countries for their own circumstances.

---

98 Cookson, *Age of Machinery*, 164–8; inf. John Suter. See also Scrivenor, *History of the Iron Trade*, 95, in which Wibsey Moor, noted with two furnaces in 1796, denotes Low Moor.

Because hard technologies are only a part of this advance, the broader perspective of industrial practice is not incidental. The 'supportive', the organizational and more local features can all be highly significant. Localities and landscapes mattered, in what they demanded as well as what they offered to innovators. What had that site been earlier? What was the historical basis of innovation, what were the foundations being built upon? How had water channels, transport systems, associated commerce, and industries come into being? How far did an innovation rest on a catalyst of supportive technologies, on close access to coal- and iron-mining, power sources, ancillary products such as improved types of rope for haulage, or breakthroughs in technical understanding? Technique and working method, and the experience and skills which underpinned them, were vital in solving problems.

Nor should the small-scale and messy elements of getting along in ironworking be ignored. Innovation was not confined to narrow channels. There were pleasingly mundane elements in how the iron trade progressed. From Darby's iron cooking pot, the product which enriched him sufficiently to enable countless further technical experiments, to the pioneering whitesmith Hattersley's willingness to take any kind of forge work in Keighley – mending spades, clog irons, or the church gates – in order to pay the bills while developing lines in precision components, this movement was indebted to the quite ordinary.[99] What is more, very many successful businesses associated with the iron trade never moved beyond a traditional mould.

The power of Angerstein is that he reflected much of that complexity, in his own time – the large and the small on the brink of great change. His observations are concerned above all with what real people were doing in real time, and reveal that actuality: inventiveness was widespread, and so was practical knowledge. Technological know-how spread quickly through well-established networks of forgemen and furnace workers, many of these employed by the famous combines, the Foleys of the west Midlands and Spencer associates in northern England.[100] Innovation and initiative in ironmaking were most associated with managers and junior partners, their trials often in association with sections of the workforce. Syndicate owners, many by then thoroughly gentrified, were losing such significance.[101] Their gradual withdrawal over the mid-century decades opened doors to a new style of active ownership. Meanwhile, ironworkers continued their migrations, mingled, and intermarried, forging new channels for knowledge and information.[102]

---

99 WYAS Bradford, 32D83/2/1; Cookson, *Age of Machinery*, 94.
100 Evans and Rydén (ed.), *Industrial Revolution*, 17–18.
101 Ashton, *Iron and Steel*, 218–19.
102 Innovating communities are further explored in ch. 7.

3

Enlightened engineering

Untersuchen was ist, und nicht was behagt.
(Investigate what is, and not what pleases.)

Johann Wolfgang von Goethe, *Der Versuch als Vermittler von Objekt und Subjekt* (1792)

The interface between industrial endeavour and science in this period is a point of controversy. Finding certainties is difficult, where the ground was shifting beneath: it was not only institutions and practices in flux, but so too the basics of knowledge and understanding.

There is potential for fresh insights. How, as much as can be determined, did science and industry relate in significant areas of industrial development? Given the many limitations of enlightenment science, could its influence have been as significant as is sometimes claimed in creating technologies and sustaining industrialization before 1800?

In contrast, the importance of craft, skill, and collaborative endeavour to industrial change has often been downplayed. Social class had a substantial impact on opportunities and working relationships, and that too, as well as how social environments adjusted to accommodate new realities, is overlooked. Historians have at times misconstrued relationships between science and technological processes, and misunderstood local contexts and terminology.[1]

How closely the two worlds, of natural philosophy and industry, operated and co-operated is of very great significance. It is evident that mathematical theory (which was well established and potentially relevant) made few inroads into mechanical workshops during machine-making's foundational period.[2] But ironmaking was a different kind of industry, to which science appears to have had greater connection and may well have contributed. Unlike machine-making, it was already well established, a heavily capitalized key sector and

---

1 For critiques, see Wakefield, 'Butterfield's Nightmare', esp. 240–4; Ashworth, *Industrial Revolution*, 145, 244–5. Specific examples of misunderstood evidence concern the roles and backgrounds of Gott, Marshall, and M'Connel & Kennedy: see below, pp. 69–70 and fn 25.
2 Cookson, *Age of Machinery*, 174–81.

promoted by the educated and influential. Of all the industrial activities where science might find practical purpose, ironmaking appears a prime candidate. And so it was, for we know that academic scientists were directly recruited to address technical problems in certain progressive eighteenth-century ironworks.

But how typical were such links, and what more is to be discovered? How far did science inform and support industry, providing theoretical tools which enabled manufacture to flourish and to develop technologically? Was this restricted to an observational role, a matter of describing and disseminating industrial procedures? How much analysis was involved? What follows is an evidence-driven attempt to clarify relationships between these various endeavours.

There are unexplained disconnects in the paths of chemistry, metallurgy, and other sciences, compared with the progress of the industries with which they were most associated. So where is 'science' (as then understood) to be found, and can the nature of its dealings with industry be established? This is not really a problem, as 'science' did not hide itself away, and its practice rested upon social networks. The important industries in this debate, ironmaking and metal trades, like chemistry and metallurgy and other nascent sciences were all experiencing upheaval. Indeed, disruptions, breakthroughs, and confusions were common to many trades and activities at this time.

Some of the new stories to be told are about disappointment. Lack of success can be instructive. It is easy to discount potentially interesting individuals because they made mistakes, or did not fulfil an evident promise. But explaining those experiences is part of the story of industrial development. Awkward episodes, or diversions from settled narratives, sometimes carry significance. So too might individuals, seemingly peripheral names with little to show for themselves, who turn out to have well-documented links into networks which afterwards came to matter a very great deal.

Consider William Lewis, who merits a whole section of this chapter. Now little remembered, Lewis was at the height of his powers as an experimental chemist during the 1750s and 1760s. He had unusually close connections with industry, at a time before chemical elements were fully understood. Lewis made errors, and has been rather sidelined from (what tends to be viewed as) the onward march of science. This has to be challenged, for, while Lewis may have been wrong, at times significantly so, that ought not to be read as 'therefore he does not matter'. For some decades this man stood at the vanguard of science, and gave significant service as an industrial chemist. He was at work in those most interesting mid-century decades. This much he had in common with (on the industrial side) Angerstein, and while neither is now well known, for a time both these men carried a torch for progress.[3]

---

3   In this paragraph (and many others in this chapter), several of the designations used – for instance, 'experimental chemist' – are anachronistic. There is no sensible alternative,

Chemists showed considerable interest in ironmaking processes. In light of their own fragility in matters theoretical, which continued beyond 1800, it is of course possible that they took from the practice of ironworking rather more than they could offer in return. Indeed, for some time it appears that industry was advising chemistry, rather than the reverse. The evidence of this rests on timings. Once again, chronology matters. Clarity demands that we 'historicize the thing'.[4]

The historian has privileged oversight, something denied to the actors themselves. Those directly involved could not perceive the whole, nor even the substantial part, of what their endeavour meant, and what it would come to mean. But we can relate people's careers to the state of contemporary technology, and to knowledge and ideas both 'practical' and 'scientific'. The social and technical concerns that weighed upon individuals in their time and place offer vital context.

Again, language is important. The eighteenth century is sufficiently close that we think we understand its meanings. But it is a foreign land, and we must take care in understanding language as contemporaries understood it. 'Science' itself is contentious. What eighteenth-century scientists thought 'science' was, is different from how later generations would define science. To assume, or to back-project a more modern understanding, is a mistake.

So what constituted a 'scientist'? William Lewis's methodical focus seems to qualify him as such, a scientist and an industrial chemist, though before such a term existed. Others in the field are more difficult to label. This applies to some leading figures, in a time when it was still possible to make important contributions across different areas of knowledge and technology. Of John Smeaton, Joseph Priestley, James Watt – all great names – none was wholly, perhaps not even primarily, a scientist. To contemporaries, they may even have seemed capricious in their shifting interests. With hindsight, individuals' purpose and direction become more apparent, and of course the drivers were not purely 'scientific' (as now understood).

Nor is the word 'arts', as the eighteenth century would understand it, quite straightforward. The *Oxford English Dictionary* includes an obsolete usage among several other definitions of 'art': 'skill in the practical application of the principles of a particular field of knowledge or learning; technical skill'. It gives as examples 'industrial, mechanic, mechanical, useful' arts. This explains the otherwise confusing title of the Society of Arts; and the meaning of 'artisan', a worker in a skilled trade, or in crafts, and only later 'one utilizing traditional

---

other than excessive use of quotation marks. On the benefits of inevitable anachronism, see Wakefield, 'Butterfield's Nightmare', 238–9.
4 Wakefield, 'Butterfield's Nightmare', 239, 236.

or non-mechanized methods'. So an eighteenth-century artisan was a type of technologist.[5]

Because class and social context mattered – defined as they then stood, within the history of their time – they matter also in this debate. Society worked to exclude, and to include. The era's fine social gradations explain much about what people knew, and why they behaved in certain ways.

While anachronism is a persistent difficulty, it points to a way forward. Wakefield's view is worth repeating.

> A history that honors the categories of the past is not a history without anachronism. It is, rather, a way to discover new things. The shape and texture of the past changes when we transform the terms in which it is written. Some things recede from view; others open themselves to us.[6]

The danger, he continues, 'is not merely that the past becomes the echo-chamber of the present; more often, and more insidiously, past echoes past: intending to write about one period, we become ensnared – often without knowing it – in the terms and categories of other times and places' – until at times, 'it is easy to forget that we are in the eighteenth century at all'.[7]

This, then, is the challenge. The specifics are of real importance, and the institution central to all this is society itself. But through an approach both open and critical, the eighteenth century opens up to us.

The blue flame

The success of the puddling process depended on workplace skills, the forgemen's strength and expertise in manipulating molten metal. Before that could happen, the worker must recognize that the time was right to start, by looking out for the blue flame which signified that carbon and oxygen were reacting.[8]

That reaction is now simply expressed, and easily understood by anyone with a basic grasp of chemistry. But what did the blue flame mean to those involved in this process in the 1780s, and before that? How did they speak of it? What did the wider world of science make of it? How was the phenomenon described and shared?

Those questions are basic to working out how science and industry related to one other – what was possible, and how in practice it was communicated. In the ironworks, as Evans shows, 'Change had to be negotiated via a workplace culture

---

5 There is hardly a page in this work that does not owe something to the OED.
6 Wakefield, 'Butterfield's Nightmare', 238.
7 Wakefield, 'Butterfield's Nightmare', 239, 241.
8 See above, pp. 48–9.

that allowed only limited linguistic and conceptual scope for innovation.'[9] To appreciate this is fundamental to writing history. Back-projecting our present state of knowledge does not help. What is more, it plants misconceptions which are difficult to shift. Even Clow and Clow, with doubtless innocent and helpful intention, when quoting Dundonald on sodium sulphate refer to H_2SO_4.[10] With this, introducing a modern chemical notation to a transcribed 1785 text, there comes a subliminal hint that people in the eighteenth century knew more than was the case. This may be obvious, but perhaps it is not.

How, then, was chemistry communicated at that time? Wootton sees anachronism as an ever-present danger because 'the language in which we write about the past is not the language of the people we are writing about'. The words 'science' and 'revolution', he says, 'are our words, not theirs'; and 'scientist' was coined only in 1833, as a useful word to express something that already existed. Scientists were active in Stuart England, some of them amateurs living on their own means, others professionals practising science in the course of paid work. The latter might be employed in universities, medicine, cartography, instrument-making, architecture, or other spheres. Serious scientists therefore existed long before the term 'scientist'. Wootton calls this 'how things precede words'.[11]

This applies equally in the closely related disciplines of chemistry and metallurgy. Carbon and iron were recognized from ancient times, though identified as elements only much later.[12] The puddling process was adopted at a time of unprecedented scientific activity, when numerous elements were first identified or confirmed as such. Oxygen, the other essential component of puddling, was isolated in 1774 by Joseph Priestley, and independently by Scheele in Sweden and Lavoisier in France. Priestley called the gas 'dephlogisticated air', relating it to the phlogiston theory developed by a German professor, Georg Ernst Stahl, in 1703. Lavoisier afterwards renamed it oxygen. Neither he nor Priestley understood exactly what they had discovered.[13]

There had been attempts to systematize chemical elements into groups. William Lewis's 'table of affinities' was the first such in English.[14] Lavoisier, in

---

9   Evans, *Labyrinth of Flames*, 75.
10 Clow and Clow, *Chemical Revolution*, 102; and see their 'Glossary of Dead Chemical Language', 618–22.
11 Wootton, *Invention of Science*, 22–32. 'Scientist' took some further decades to gain acceptance: Wakefield, 'Butterfield's Nightmare', 233; Ross, 'Scientist'.
12 See the Royal Society of Chemistry for a dated list of discoveries: https://www.rsc.org/periodic-table/history.
13 Wootton, *Invention of Science*, 86–8, 91. Wootton makes a further point about these simultaneous events having resulted in rather different discoveries: 91, also 107. For a synopsis of Priestley's position, see Schofield, 'Joseph Priestley', DNB.
14 Sivin, 'William Lewis', 67–71, and see below.

counting 33 elements, included oxygen for the first time.[15] Subsequently, John Dalton transformed the philosophy of chemistry with his chart, published in 1808, identifying known elements with symbols. Later the Swede Berzelius changed that to a letter notation, the basis of the periodic table as we now have it.[16]

With oxygen, it took around forty years to fully comprehend and recognize it for what it was.[17] Cort's method came into use just as this was happening. So the chemistry – if that means a system through which his industrial process might have been understood – was not then available. As Cardwell suggested, before you can apply science, you must have a science to apply. 'How, for example, could there be a scientific "applied chemistry" when the phlogiston theory dominated that science?'[18]

Where were the eighteenth century's chemical manufacturers in this? Some of these producers, particularly in Scotland and the west Midlands, were apparently progressive, involving scientists as advisers. Many other industries used chemicals, or had chemical reactions at the centre of their operation. For the iron industry, belonging in the latter category, this had been the case for as long as iron had been worked. A great deal had been achieved with (as a modern observer might judge it) insufficient theory. Making chemicals requires 'the existence of the necessary raw materials and sufficient theoretical knowledge for their utilization', says Clow.[19] But how to measure what is sufficient? Practice was at that point running ahead of abstraction.[20] As late as 1961 it was suggested that 'theoretical understanding of the nature of metals ... has rarely led directly to new discoveries, simply because empirical experiment had discovered them in advance of theory'. A metallurgical plant, it was said, 'is a living museum of entrancing phenomena'.[21]

However true that was in the 1960s, it indicates that those key eighteenth-century industries are best understood empirically. That does not mean that ironmaking and mechanical engineering developed independently of science. Bacon's philosophy of science, founded in the idea of discovery, was widely

---

15 See Siegfried, *From Elements to Atoms*, especially section 11, 'A Compositional Nomenclature', 183–93.
16 F. Greenaway, 'John Dalton (1766–1844)', DNB; Dalton, *New System of Chemical Philosophy*.
17 Wootton, *Invention of Science*, 87.
18 Cardwell, *Organisation of Science in England*, 16–17.
19 Clow and Clow, *Chemical Revolution*, 43, and ch. 2 *passim*.
20 'Underdetermined', defined in the OED as 'account[ing] for (a theory or phenomenon) with less than the amount of evidence needed for proof or certainty', is used by historians of science to describe this. Thanks to Barbara Hahn for her ideas.
21 Smith, 'Interaction of Science and Practice', 357. The second quotation is a comment by R.F. Mehl.

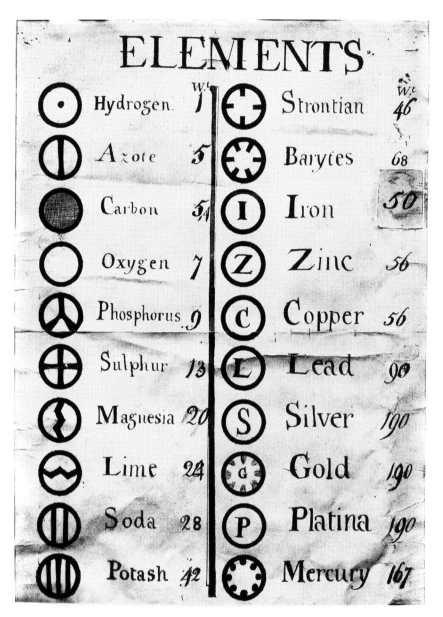

Plate 5. Dalton's symbols of the elements, 1806. Source: Wellcome Trust, WT/D/1/20/1/37/6.

accepted by eighteenth-century science practitioners.[22] But the relationship between these two practices, of science and engineering, was most visible on the ironworks shop floor, and in the mechanic's workshop.

Sharing knowledge

Before 1800, most practical transmission of knowledge about metallurgy, chemistry, or mechanics can have happened only through personal contact, given how scarce were relevant books and other published sources. Primarily, we would expect, knowledge was shared by training and example in workplaces. Few written works, technical or scientific, were available, and fewer still in English. Even if these had been on hand in the workshop, most tradesmen would have found them difficult or completely impenetrable.[23]

But a palpable eagerness for knowledge ran through the new and evolving industries. It may be that written works existed which could help inform an innovative shop floor, or benefit an educated ironmaster seeking to improve production methods. Or if not books, then there was another natural route to follow when engaging in the unfamiliar: to consult others with specialist knowledge, and make notes.

The enquirers, the keepers of notebooks, are mainly of that class which had some means and education. A few were wealthy, and others later became so through these very endeavours. Some were investors ready to join as partners with technical men. This might be a practical as much as a scientific exercise, as the world became a more complicated place.[24]

A distinction is needed between records which show potentially useful knowledge transferring between individuals; and the private notes used within businesses to log trials, whether or not successful, on the shop floor. Examples of the latter, recorded for internal use, are John Marshall's flax-spinning trials with Matthew Murray; and the notes of Benjamin Gott's foreman and others in the routine business of expanding their range of dyes.[25] In both instances,

---

22 Wootton, *Invention of Science*, 91, 107.
23 Education levels are discussed in following chapters.
24 See pp. 15–16.
25 Rimmer, *Marshalls of Leeds*, 23–4; A. Yewdall, 'The Bean Ing Mill Notebook', in Crump (ed.), *Leeds Woollen Industry*, 272–307, referring to Brotherton SC, MS 193/117A. Clow and Clow, *Chemical Revolution*, 220–3, rely on Crump's synopsis and are wide of the mark. But neither Rimmer nor Crump supports the idea of 'science' being applied to technological progress in textiles, as repeatedly suggested by Jacob, *First Knowledge Economy*, ch. 4; 'Mechanical Science on the Factory Floor', etc. Benjamin Gott had an interest in chemical classification, and arranged to have his son instructed (from c. 1808) in chemistry. The notebook, though, is an eighty-page compendium of technical and commercial information,

the expertise of employees was deployed, but it was skill- and experience-based. Later, Marshall, faced with a linen-bleaching bottleneck in his operation, invested considerably in industrial trials with chemical bleach. This was in 1798, a decade after Berthollet discovered chlorine's properties and Watt applied it to linen-bleaching; and a similar period since Marshall's mechanical success with Murray. Marshall probed how Lancashire cotton spinners sourced and used chemical bleach, and read some of the available scientific literature. He explored what could be achieved in practice, and what it would cost, including products available commercially. He could not recapture that success enjoyed with Murray. Quickly it became clear that the blend of economics and quality he sought was out of reach. Instead he adopted a number of ad hoc systems.[26] The contingency suited Marshall's own works, and was emphatically not experimental science.

Where, then, did science sit in this? Theoretical chemistry and metallurgy, as well as mathematics and physics, existed as streams running alongside the maturing flow of industrial practice. There would be some later convergence, though not a confluence. Their union was never absolute. Meanwhile science appeared to be struggling to find its practical utility. There was, too, a level of difference between the information (whether written or word of mouth) available to the new industry, machine-making, compared with the older one. Ironworking, having been around since prehistory, had a well-developed body of empirical knowledge, epitomized by the 'blue flame'. Practices used in ironmaking and mining were recorded and published from the early sixteenth century. Some of these works were of high significance, and predated – by some margin – the dissemination of Francis Bacon's ideas early in the seventeenth century.

This earlier empiricism was hard at work by the 1520s. The sixteenth century was, in the view of Long, in both volume and originality the great age of mining and metallurgical literature. She identifies three types of publication within this canon, though with considerable overlap: 'recipe' books; alchemy; and mining and metallurgical treatises that were intelligible to outsiders, the first of these written before 1523.[27] Long is interested in this period's new openness, which did not, she emphasizes, necessarily extend to a wide public dissemination of knowledge. Indeed, the writings appeared as books with short print runs,

---

in the random and untidy handwriting of many different individuals over the course of three decades. It shows ingredients and methods for a range of cloth processes, but no one reading it could possibly claim it as 'science', in any sense.
26 Rimmer, *Marshalls of Leeds*, 50–3. The link with 'science' made by Musson and Robinson, *Science and Technology*, 329–32, is not supported by Rimmer, the source they quote. See also p. 90.
27 Long, 'Openness of Knowledge', 320–1. Swedenborg's mystical treatise on iron, though mid-eighteenth century, may yet qualify as 'alchemy'.

or as hand-copied manuscripts.[28] But for Bacon, and for seventeenth-century scientific thinkers more generally, they seem to have had significant influence. Bacon, called the father of empiricism, saw science as something that 'does not slip away into sublime and subtle speculations, but is applied effectively to relieve the inconveniences of the human condition'.[29] Bacon's idea for histories of all trades, a scheme picked up by Robert Boyle for the Royal Society in the 1660s, was evidently stimulated by those works of the previous century. But Bacon had not referenced them himself, his rule being 'Never cite an author except in a matter of doubtful credit.'[30]

The best known of those sixteenth-century metallurgical treatises was *De re metallica* by Georgius Agricola, which Long places in the category of formal and outward-looking, though it was also indebted to both 'recipe' and 'alchemy' writings. *De re metallica* explained alternative methods of producing and working iron, and was illustrated with woodcuts showing the different processes.[31] Agricola was certainly influential. The editors of a new translation in 1950 made no apology for its reissue: 'During 180 years it was not superseded as the text-book and guide to miners and metallurgists, for until Schlüter's great work on metallurgy in 1738 it had no equal.' Agricola's work had passed through ten editions in three languages 'at a period when the printing of such a volume was no ordinary undertaking'. This, they said, 'is a record that no other volume upon the same subjects has equalled since'.[32]

So here are several indications about how industrial technologies were shared. These points remained valid into, though not too far into, the nineteenth century. Published works were scarce, and in some cases were not superseded for decades or even centuries. Their content tended to be descriptive and short on analysis. As innovation quickened through the eighteenth century, they soon became outmoded. The dynamism and immediacy of industrial change were not, could not be, captured. Abraham Rees, in his thirty-nine-volume *Cyclopaedia*, published between 1802 and 1819 at a high point of British industrial innovation, was very much in this school of recording settled technologies.[33] Could he have done otherwise? The difficulty was that industry's most interesting machines and procedures were invariably works in progress, and that progress was constant, happening in workshops and foundries across the nation.

---

28 Long, 'Openness of Knowledge', 352–3.
29 Wootton, *Invention of Science*, 476–7.
30 Long, 'Openness of Knowledge', 354–5; Ashworth, *Industrial Revolution*, 145–6.
31 Long, 'Openness of Knowledge', 322, 334–40; Long, 'Of Mining, Smelting and Printing', 97–101.
32 H.C. and L.H. Hoover (ed. and trans.), *Georgius Agricola*, ii–iii.
33 Cookson, *Age of Machinery*, 182.

This explains why visitors, British and foreign, were drawn to industrial regions to observe and discuss the latest shop-floor techniques and equipment. While Angerstein stood out in his depth of experience, many others arrived who were keen to learn at first hand things which were not discoverable from books. The more serious observers were people of intelligence, including scientists and others with specialist industrial knowledge. Yet even among this group, there was no template.

In explaining the Cranage process, 'making bar iron without wood charcoal', Richard Reynolds – managing partner at Coalbrookdale – showed the limitations of this technical discourse. He had shared the view of most or all ironmasters, that this was impossible, 'because the vegetable salts in the charcoal being an alkali acted as an absorbent to the sulphur of the iron, which occasions the red-short quality of the iron, and pit coal, abounding with sulphur, would increase it'. Industrial trials in 1766 suggested that it might be achieved, but in practice the project largely failed. It rested heavily on manual skills, but whatever the shortcomings, it seems they could not then be substantiated.[34]

The language in which this was expressed differs from the language of science as we now know it. Elements and compounds were referred to by names which became obsolete after 1800.[35] 'Recipes' often lacked precision when it came to quantity and method. Could it have been otherwise? There was still no certainty about elements, even with the surge of discovery when important new gases were identified late in the century. Nor was there a settled framework through which to express what was known of chemistry.

The contrast between the Swedish contemporaries Angerstein, apparently uninterested in science *per se*, and Rinman, the outstanding metallurgist infused with scientific rigour, has been noted.[36] Yet Angerstein provides an outstanding model of observational practice. He measured, he drew – and the drawings are exceptionally good – he provided some history and context, and he interviewed owners, foremen, and workers to collect information about materials, quantities, processes, and methods. He was highly attentive to the economic aspects of the iron trade, and of the other industries he inspected. He worked in the spirit of enlightenment improvement, looking to learn and to find ways of building integrated ironworking enterprises, from mining right through to marketing. It was a systematization of experience, with a close eye on economics. But while Angerstein is the ideal of a rational and informed industrialist, he was not a scientist.

---

34 Raistrick, *Quakers in Science and Industry*, 135–6; Hayman, 'Shropshire wrought-iron industry', 10–13, 55–61; and see Hayman, *Ironmaking*, 47. Reynolds' letter was to Thomas Goldney, 25 Apr. 1766. 'Red-short' here was an undesirable property of iron under sustained hammering.
35 Clow and Clow, *Chemical Revolution*, 618–22.
36 See pp. 31, 34.

William Lewis and the practice of science

William Lewis (1708–81) is a neglected giant, the 'forgotten man of eighteenth-century chemistry'. Admitted Fellow of the Royal Society (FRS) at an early age, he was an industrial chemist before such a thing was recognized, esteemed as a fine experimentalist, distinguished and honoured in his lifetime, and carried considerable influence. Lewis is highlighted here because he was a pioneer, campaigning through his own work to engage science, particularly chemistry, in the cause of improving industrial efficiency, especially in ironmaking.[37]

Lewis's unconventional approach partly accounts for the subsequent lack of notice. He knowingly chose, as Sivin explains, a career which was practical rather than academic. That is not to imply that he lacked rigour: Lewis was careful and thorough, and was a collector and, through his loyal assistant Alexander Chisholm, translator of the works of German and other chemical theoreticians. He used theory judiciously. Publishing his pioneering version of a table of chemical affinities in 1753, Lewis was clear that this was shared only for its practical value. He did not even, says Sivin, see affinity as a general, quantifiable, doctrine. 'Chemical facts stand alone and are "discoverable by observation only".'[38]

Following the route taken by many of those who would subsequently engage in scientific research, Lewis trained as a physician. He continued to practise medicine while also attracting a number of aristocratic and scientific patrons for his research. He had studied at both Oxford and Cambridge universities, and began lecturing on chemistry and pharmacy when still in his twenties. He was a prolific writer and publisher throughout his life. In 1747 he created a well-equipped laboratory at his home in Surrey, where he explored new ways of analysing chemicals. Lewis was the first to recognize platinum as a distinct metal, and that interest in precious metals continued as a theme in his researches. In about 1750, he employed Chisholm, an Aberdeen graduate, to search and translate scientific literature. This arrangement endured until Lewis's death.[39]

What, then, of Lewis's industrial connection? Certainly some manufacturers, including ironmasters, sought his advice. One of his major achievements came out of a commission from the Society for the Encouragement of Arts, Manufactures and Commerce to analyse American potashes. The procedure

---

37 Gibbs, 'William Lewis, M.B., F.R.S', is the most complete biography of Lewis; see also Sivin, 'William Lewis', from whom the quotation is taken (pp. 63–4); Gibbs, 'William Lewis and Platina' and 'Notebook of William Lewis and Alexander Chisholm'; Schubert, *History of the British Iron and Steel Industry, passim*.
38 Sivin, 'William Lewis', 75 and *passim*.
39 Gibbs, 'William Lewis, M.B., F.R.S'; Sivin, 'William Lewis'; Gibbs, 'Notebook of William Lewis'; Page, 'William Lewis (1708–81)', DNB.

developed by Lewis enabled large-scale potash production in Massachusetts, the method still used more than a century later.[40]

Lewis died leaving a collection of more than 200 bound volumes, the records of his investigations. The collection was broken up for sale by auction, when many fetched high prices. But it was scattered, and little of it can be located.

Lewis's two most significant published works shed light on how he saw his relationship with industry.[41] He is explicit about his own interests, and explains his belief of what is possible when science and industry unite in common endeavour. The first volume, appearing in 1748 shortly after he commissioned the laboratory, was *Proposals for Printing, by Subscription, the Philosophical Commerce of Arts*. The second, *Commercium Philosophico-Technicum: the Philosophical Commerce of Arts: Designed as an Attempt to Improve Arts, Trades, and Manufactures*, was the outcome of the first, issued in two parts in 1763 and 1765. *Proposals* is a manifesto, while the preface to *Philosophical Commerce of Arts* shows how Lewis struggled with the scope of that original ambition and had been persuaded to modify his scheme. Despite that, or perhaps because of it, the 650 pages of those volumes, appearing after fifteen years of research, deliver range and authority.

The prospectus laid out by Lewis in *Proposals* aimed to attract subscribers to meet publication costs. Exactly who did subscribe is not known, for (though promised in *Proposals*) it does not seem that the list was published. The works of 1763–5 were dedicated to the king, George III, who had attended some of Lewis's lectures and demonstrations. Royal endorsement, along with the patronage of the Earl of Yarmouth, the Duke of Northumberland and the celebrated and wide-ranging scientist Dr Stephen Hales – all of this would have helped gather support from moneyed gentlemen. *Proposals* suggests, though, that while its main target was finance, Lewis was hoping for more: contributions of understanding and experience from industry were key to his method.

Lewis's pledges in *Proposals* underscore his interest in some larger possibility, of science and trade as symbiotic, working to improve economy and efficiency in business. In truth, his account here does focus largely on the potential benefits to industry. But of course this was a flotation, presenting inducements to invest in his vision. He was rather less expansive about the scientific opportunities he would have, to pursue empirical investigations in workplace settings, which was clearly integral to the plan. How he balanced the priorities, and how those shifted over the course of the project, is a matter of speculation. This was, after all, a package, and Lewis's interests ranged widely.

He promised new experiments and analysis that would determine the 'properties of many substances', including ones 'not as yet introduced into

---

40 Sivin, 'William Lewis', 77.
41 Gibbs compiled a full bibliography of Lewis's publications: 'William Lewis, M.B., F.R.S'.

arts or trades'. He would be looking 'to substitute, for scarce and dear ingredients, such as are cheap and easily procurable', including home-produced parts in place of foreign ones. Imported products were to be examined, with a view to emulating them. He offered what were in essence audits of different trades, testing and defining what worked in practice, and how other processes might be improved. He sought to eliminate 'tedious' tasks so as to 'spar[e] as much as possible the labour of the operator'; new methods would aim not to diminish the products and 'tend rather to render them more beautiful, durable, and better fitted for the uses of life'.

What is to be read from this? His economic perspective meant that industrial efficiency was important to Lewis. His plan also reflects the preoccupations of social circles situated higher, or much higher, than his own circumstances, as son of a prosperous brewer in Richmond, Surrey. The agenda for Lewis's investigations and improvements included the colours of oil paints (and of varnish and house paints), porcelain manufacture, and stronger window glass. The refining of gold, silver, and other precious metals was one specific area of interest.

Yet Lewis's strategy also turned to far less glamorous arenas. His interest in colour was rooted in dyeing, in textiles and printing besides more general usages. He planned to investigate brewing, soap manufacture, potashes, salts, and sugars. Also on his long list was an agenda to explore the properties of metals, in particular iron and steel. He would study 'the business of fluxes', probe 'the common methods of smelting iron ore', attempt 'to increase the malleability of iron, both cold and hot'. And so on, through types of cast iron and steel, melting, hardening, and tempering to change the brittleness and other qualities.

An ironworks, it should be recalled, the 'living museum of entrancing phenomena', offered scientists an experimental paradise.[42] Inevitably this must, and did, interest Lewis. And his programme of work illustrates, in content and approach, just how much his eye remained steadily focused on science as he knew it. The suggestion of platinum as an element in its own right emerged from this project, from his investigations into precious metals in the early 1750s. His promise to produce tables of salts and other substances took shape in a table of affinities, published in 1753 in *The New Dispensatory*.[43]

From the start, Lewis's investigation was framed by scientific method. His descriptions used consistent standard weights and measures. His experiments were to be transparent, communicating 'the whole, without reserve', and revisiting anything which turned out not to be reproducible. He hoped that some of his readers would try to emulate his findings for themselves, in a workshop or even at home. He was above all a teacher, and by all accounts

---

42 See above, p. 67.
43 Lewis, *The New Dispensatory*, Pt 1, ch. 2, 28–9.

an excellent one. The equipment he recommended was simple, and the first section of the book would contain detailed instructions in how to construct 'a small apparatus', a furnace suitable for experimentation.[44] This project was rooted in empiricism.

Yet the preface to *Philosophical Commerce of Arts* launches immediately into a discussion of where the boundary lay between physics and chemistry. But this, says Sivin, was probably the only statement Lewis ever made about the theoretical basis of chemical action. Lewis describes the difference between natural (mechanical) philosophy in terms of *attraction*, of masses and movement, and all 'reducible to mathematical calculation'. Chemistry is atomic, separating and blending, 'governed by laws of another order', *affinity*.[45] Affinity in a context of science is defined as a resemblance in structure, properties, or composition, so, in Lewis's case, between different mineral substances.[46] Here in his preface, and (previously published) in his own amended version of a table of affinities, are contained all that Lewis wrote about the doctrine of the affinity of bodies. Sivin suggests that Lewis had little interest in developing the theory. To him, the importance of the table rested in its practical value as an aid to experimentation.[47]

In the preface to the later work, Lewis explains that the original plan had changed. Having taken advice from unidentified 'friends', he modified the presentation of his findings. He had noticed that many 'arts' – meaning trades and industries – had much in common with others, yet no body of shared technical knowledge existed. Little or no exchange took place between, for instance, dyers working in different branches of textiles, or between those groups and others working in colours – glass and porcelain manufacture, or the finer arts. The same applied to the numerous skilled trades working iron for an array of purposes. Lewis had therefore chosen an arrangement which made this common ground more obvious, dividing it by 'arts'.

The chemistry practised by Lewis was not then in shape to respond in a scientific sense to industry. That is not to say that he himself had little to offer. Indeed, he made great practical advances through observation and reflection, and some of his insights appear to have been particularly helpful to the iron industry. Latching on to something promoted by the Académie des Sciences, a project launched in France almost a century earlier, Lewis advanced the idea of a 'history of arts'. This meant collecting a corpus of principles common to (for example) all iron trades, and sharing this, applying it for wider benefit to see where things could be done better. The 'history' aimed at a collective wisdom, not so much the specialisms of particular trades. This was not ephemeral, and

---

44 Lewis, *Proposals, passim*; Sivin, 'William Lewis', 67–73, 85.
45 Sivin, 'William Lewis', 74–7; Lewis, *Philosophical Commerce of Arts*, iii–v.
46 OED. The earliest examples presented of this usage are dated 1671 and 1718.
47 Sivin, 'William Lewis', 68–73.

it would be rigorous. The enquiries were, said Lewis, 'founded on the invariable properties of matter', and 'a principal part of the work' comprised experiments and research which would deliver practical improvements. He invited feedback, and as promised the book opens with directions for a cheap and easily managed set of furnaces working under a common chimney. This, he suggested, could be in a domestic setting – 'in the middle of a room without offence'. For 'one of the principal obstacles to the prosecution of chemical enquiries has hitherto been the want of a proper apparatus'.[48]

Nonetheless the *Philosophical Commerce of Arts* drew heavily on secondary sources. Lewis's observations on 'the principal machines used for blowing air into furnaces by a fall of water' trace the evolution of this technique from the earliest, Roman, times. His background investigation derives entirely from printed accounts, and that applies to the modern sites as much as to the ancient. With his most recent examples, one described at Lead Hills (Lanarkshire) in 1745, and several in France reported by Réaumur in 1762, there is no indication that he or Chisholm visited the sites. The surviving evidence suggests that they did not travel, or at least not routinely.[49] So the initial work was a history, of what had been, and what was. Lewis's own experiments proceeded from this base, and he recounted his investigations in some detail. He was also informed by, and took close account of, data and discoveries from other men of science. These included Lewis's mentor and patron Stephen Hales, and Desaguliers and others.[50]

Beyond doubt, the loss of those hundreds of volumes detailing Lewis's research has been a blow to developing a better understanding of what he and Chisholm achieved, and how they worked. With those records unavailable, how much can be known of his industrial contacts? The key question is perhaps 'Who was asking for advice from whom?' Was this about seeking an expert opinion from the scientist? No, because enough is known about Lewis's work to see that this information was two-way traffic, and the person most consistently chasing data seems to have been the man himself.

Missing sources are a frustration, but not fatal to finding answers. Gibbs's admirable biography of Lewis from 1952 addresses earlier misunderstandings, filling out the life story and tracing influences.[51] Regarding his working methods and contacts, something similar can be performed by assembling

---

48   Lewis, *Philosophical Commerce of Arts*, iii–xviii; Sivin, 'William Lewis', 84. Regarding the furnace in the drawing room, note that Lewis was unmarried.
49   Lewis, *Philosophical Commerce of Arts*, 270–9; Réaumur in *L'Art des Forges et Fourneaux à Fer* (Paris, 1762), a translation from Latin by M. Bouchu, and to which Swedenborg also contributed.
50   Lewis, *Philosophical Commerce of Arts*, 279–314. John Theophilus Desaguliers is discussed at greater length in ch. 4.
51   Gibbs, 'William Lewis, M.B., F.R.S'.

what is retrievable of Lewis and Chisholm's records. Up until the time that *Philosophical Commerce of Arts* was published, the train of Lewis's interest is there to be seen. But for the following period, from 1765 until his death in 1781, the time when Lewis paid increasing attention to the science of ironmaking, there is no equivalent summary of his investigations and conclusions.

Gibbs drew on a small but significant set of letters written to Lewis over many years. Correspondents included the Cockshutts of Wortley and Seamer forges in the 1770s, and there were early drawings and a description of John Wilkinson's Watt engine.[52] Gibbs also used the Wedgwood archive, which holds a small section of Chisholm's 'commonplace' records – there was no sign of the laboratory notebooks thought to have been bought by Josiah Wedgwood after Lewis died. The Chisholm fragments confirm that Lewis continued to range widely in his interests, and also that his focus on ironmaking intensified. They include a list of specimens gathered on his and Chisholm's excursion to a number of Midlands and south Yorkshire ironworks in 1768.[53] Overall, Gibbs was able to show the range of Lewis's contacts, academics and industrialists and others, home and abroad, whom he frequently approached for advice and information. This network responded with data for Lewis's studies. All was systematically recorded, mainly by Chisholm: transcribing extracts from technical literature, making notes of their experiments, writing fair copies of Lewis's papers intended for publication.

Unknown to Gibbs, though – indeed barely known at all, except to Schubert who extensively quoted from it – late-period Lewis left behind a work which is equivalent to his mid-life *Philosophical Commerce of Arts*. In Cardiff is a six-volume 'Mineral and Chemical History of Iron', unfinished on Lewis's death and never published.[54] This offers firm evidence of the scientist's focus in his later years. In volumes IV and V (at least), a wealth of information was collected and discussed, and further questions raised, concerning iron and steel production, ironstone- and coal-mining and the uses of different grades of those minerals, furnace types, and methods of wire-drawing and file-making. Lewis was especially interested in the Barnsley area, including Wortley, and parts of Lancashire.[55]

52 Gibbs, 'William Lewis, M.B., F.R.S', 145–7. The collection is in the BL: Egerton MS 1941 (1735–76).
53 V&A Collection at World of Wedgwood, Collect. Chem. Experiments 39-28405 (1771–6) and 39-28406 (1768); and see Gibbs, 'Notebook of William Lewis and Alexander Chisholm'. Chisholm (1723–1805) worked for Josiah Wedgwood as scientific adviser and secretary from 1781, and served the family for the rest of his life. The commonplace books were later acquired by the Wedgwoods.
54 Cardiff Lib. Manuscript Collection, MS 3.250; Schubert, *History of the British Iron and Steel Industry*, esp. 231 fn.
55 Most of this summary is from Schubert, as circumstances in 2022 prevented more thorough searches of the Cardiff set. It seems that the volumes (mistakenly catalogued

Queries — answered by Mr. Cockshutt.

Is the ore found at day, or dug out of mines.

The ore appears at day in the same manner as coals do, but none used but what is dug out of mines.

To what depth has it been worked.

The deepest of our mines are about 26 yards.

Is there any water in the mine.

It is nothing good for till the water remains amongst it.

Has the water any particular qualities, as a strong inony taste.

It has an inony taste, but not strong enough to make it disagreeable to many people.

Is the ore richer in proportion as it lies deeper.

The deeper the ore is found, it is always the stronger & more dogged in the working.

Does it lie in solid measures, with straggling balls over each.

There is always two measures or beds of solid ore, about a yard distant from each other, with the balls above the uppermost bed.

What is the thickness of the measures.

The uppermost from 2 to 4 inches thick, the lower from 4 to 10 inches.

What are the substances found chiefly intermixed with the ore or in its neighbourhood — limestone? — or Gritt? — or Clay? — or Gravel? — or Coals?

We have no limestone in this neighbourhood. From the surface to about 14 yards we meet with earth, & stones of the delf kind; sometimes the blue shale (which turns to clay by being exposed to the weather) in beds of a foot & a half thick, & then earth & stones again in an uncertain manner for about 14 yards, & then begins what the workmen call the rust; it is a vein of very dark grey and excessive hard stone about a foot or 15 inches thick; this is the

Plate 6. William Lewis's method shows in questions he posed to the younger John Cockshutt. These, with Cockshutt's responses, were recorded by Lewis's assistant, Alexander Chisholm, illustrated here from the volumes now in Cardiff. More of Lewis and Chisholm's exchanges can be found in World of Wedgwood (Collect Chem. Experiments 39-28405, 1771, and 39-28406, 1768). Source: Cardiff Central Lib., MS 3/250, vol. IV, f. 505.

Gibbs had managed to work out some of this from other sources. He shows Lewis in the mid-1770s seeking to identify the qualities of, and differences between, types of cast iron. Lewis's assistant Chisholm wrote to various associates, seeking information about how cast-iron items were manufactured. He asked that his detailed list of queries be put directly to workmen, and requested specimens of grey and white pig irons which had been produced from the same ore. Chisholm was frustrated – 'Our accounts of the coak iron works are very imperfect' – by the inconsistent data provided by different ironworks. It was not at all clear whether variations occurred because of different conditions in the plant, or in the mix of materials.[56]

It was during the early phases of this investigation that Lewis and Chisholm embarked on that single verified fact-finding trip, in 1768. This was during the 'potting' period, when strenuous efforts were in progress to find viable ways of producing bar-iron with mineral fuel. Their mineral-collecting included many samples of iron in various states of treatment – sometimes (as the pair happened to pass ironworks in Sheffield) scrap from the roadside. At Staveley, near Chesterfield, once part of the Spencer empire, the furnace was out of commission and they gained full access to measure, sample, and observe. Their main objectives, though, lay elsewhere, at Wortley and Coalbrookdale.

In those six under-appreciated volumes in Cardiff, Schubert discovered substantial evidence that Lewis pursued similar investigations into Lancashire ironworks, though perhaps only by correspondence – nothing suggests a visit. But Lewis had particular reason to break his usual habits in travelling to see the Cockshutts and the Darbys. Both were known for innovative approaches. Both had insight into the stamping and potting process. The Cockshutts were also of an unusually (for ironmasters) scientific tendency, and already acquainted with Lewis.

The younger John Cockshutt, then in charge at Wortley, knew Lewis well, from his time as London representative of the forge. Both were members of the Society of Arts. John II was educated and well informed, as demonstrated by his request to borrow *Traité de Forges et de Fourneaux* from Lewis in 1763, a new translation from Latin into French.[57] His brother James, who guided the visitors, was later a noteworthy experimentalist with an exceptional library, and elected FRS in 1804. The trials then underway at Wortley resulted in the process patented by John II in 1771. It was an attempt to refine the stamping and potting processes patented by the Wood brothers in 1761 and 1763, as Chisholm's notes confirm. Lewis was probably acquainted with Charles Wood,

---

as 19th-c.) reached the library in the 1950s – after Gibbs, but before Schubert published in 1957. They had previously been in the Royal Artillery Regiment Library, perhaps through a Bute connection.

56   Gibbs, 'Notebook of William Lewis', esp. 204.
57   BL, Egerton MS 1941 fols 6–7. See also Cookson, 'Wortley Forge', 67.

and would certainly know that Wood had brought the first platinum to Europe. Wood's brother-in-law and later partner, William Brownrigg FRS, had reported on this and passed the sample of platinum to the Royal Society in 1750.[58]

Lewis and Chisholm then travelled to the Darby works at Ketley, near Coalbrookdale, scene of the second Abraham Darby's organizational revolution. After his death five years earlier, the interim managing partner, Richard Reynolds, had supported the Cranage initiative and patented it in 1766.[59] There, the visitors gathered more samples and observations.

Among notes in the Wedgwood collection, Gibbs located a translation which offers an insight into Lewis and Chisholm's approach. Perhaps this stretches a point, and was never intended to apply to the public transmission of scientific or technological knowledge. It may have been nothing more than a reflection on how Germans saw the British. It does, though, chime with Lewis's expressed belief that sharing ideas with others is a route to better understanding and greater efficiency. It suggests that the kind of scientific and industrial collaboration in which he habitually engaged was beneficial to innovation. At least Lewis was sufficiently impressed by this excerpt from Johann Jacob Reinhard's *Vermischte Schriften* (1760–7) that he had Chisholm translate and store it:

> I have followed the practice of the English, who make known by the press such of their projects as they think may be of publick utility, and thus give every one an opportunity of pointing out their faults and advantages, of improving, extending or contradicting them ... By this means the general good must necessarily be promoted.[60]

To work collaboratively is possible only if others travel the same path. William Lewis's dealings show his active co-operation, through shared information, with associates in learned societies and industry. For him, chemical exploration was a vocation as well as a practical exercise. Other practitioners were more focused on overarching theories, or, less nobly, on building an academic career. Or they were chasing solutions to solve specific industrial problems, or perhaps viewed science simply as a stimulating and respectable pastime for an intelligent gentleman. The institutions supporting scientific activity brought together many interests.

While this was a lively mix, how effective was the world of science – here chemistry and metallurgy – in helping industry along? Lewis's records, or

---

58 Cookson, 'Wortley Forge', 66–8; Gibbs, 'Notebook of William Lewis'; 'William Lewis and Platina', 68; Hunt, 'First Experiments on Platinum'; for Wood, see also pp. 46–7.
59 Gibbs, 'Notebook of William Lewis', 208–9, 217; see pp. 47–8, 72.
60 Gibbs, 'Notebook of William Lewis', 205, quoting Chisholm's translation from Johann Jacob Reinhard, *Vermischte Schriften* (1760–7). Reinhard (1714–72) was a constitutional lawyer and university teacher with interests in statistics and economics.

what can be discovered of them, show considerable interaction with the iron industry, with a focus on collecting data for his own work. The renowned studies of potash were of immense significance to production, based on new methods and measurements, and a hydrometer of his own invention.[61] But in iron-processing, while his observations were doubtless useful, Lewis had less value as a consultant. He had lived and worked within a framework of chemical theory that would soon be decisively replaced. He died before puddling solved the ironworking problem that he and many others had spent decades pondering. Ultimately it was overcome through workshop practice, the 'blue flame' empiricism found on the shop floor rather than in the laboratory, while science ran to understand.

The scientific establishment

William Lewis, so eminent a chemist, so directed and meticulous, was not typical in choosing a path that was also practical and industry-focused. But in his own time, a working life spanning the 1740s to 1770s, the typical scientist was an elusive character. The discipline of chemistry was feeling its way, settling fundamental principles. Lewis and his contemporaries were part of that process, as they built and tested knowledge. In straying across other fields of learning they (perhaps unknowingly) created boundaries of their own. More definitive lines were drawn by the following generations.

Chronology matters here. The scientific interests and methods of those at work in the mid-eighteenth century, often ranging far and wide, make them difficult to categorize. Some well-known men of science apparently allied themselves with industrialists, and new institutions promoted modernization. But the society memberships and personal connections, the nature of interactions – were these intellectual, professional, social? Most likely the value for many people lay in a mix of those three. But that is still some distance from proving how, and to what useful extent, personal contacts in the real world delivered scientific contributions which benefited industry.

One generality which turns out to be well founded is that, in actively progressing both industry and science, Scots were some distance ahead of their near neighbours. The Board of Manufactures, established in 1727 in the wake of political union with England, was charged with advancing Scottish industry in balance with its English counterparts.[62] The Board's trustees encouraged a

---

61 Sivin, 'William Lewis', 77.
62 See ch. 7, 'Galaxy of Scotsmen', for more. Clow and Clow, *Chemical Revolution*, xi–xiv, 21, 41; National Records of Scotland, NG1 (https://www.nrscotland.gov.uk/research/guides/scottish-government-records-after-1707).

range of industries, and later in the century Scottish university scientists picked up on the opportunities, to good effect.[63]

South of the border, the situation was different. The Royal Society, formal and exclusive, positioned itself apart from everyday pragmatics. It was a creation of the Restoration, founded by royal charter in 1660, and its early ethos was directed towards challenging atheism, rather than diffusing useful knowledge to industry. The social divide within its ranks, between gentlemen and technicians, was very apparent. The latter group (broadly defined as those interested in practical matters, and in sharing knowledge) spilled out into coffee houses, and to some extent towards the provincial lecture circuit. Through these channels, it is argued, the works of Newton and Boyle became more widely known.[64]

The Society for the Encouragement of Arts, Manufactures and Commerce formed in 1754. Its initiator, William Shipley, noticed the effect of cash prizes in encouraging racehorse breeding, and promoted this new group (quickly renamed the Society of Arts) in a similar mould.[65] Premiums offered by wealthy patrons would incentivize innovation. Perhaps Shipley had read Lewis's *Proposals* and been struck by the promise of connecting science with trade, to promote efficiency and economy. The Society of Arts was certainly comprehensive, embracing agriculture, commerce, and the fine arts. In its early years, society premiums supported competitions to improve windmills, but also gave prizes for paintings by the daughters of lords. William Lewis's patron Stephen Hales was a founding vice-president, and it might be expected that the industrial work of the Society of Arts would be a natural fit for Lewis himself. But Lewis did not immediately embrace the society. He joined in 1760–1, and in 1767 had the distinction of receiving the society's gold medal for his work on potashes.[66]

Who else was then interested in the practical application of science? And how was scientific knowledge, as then understood, disseminated around and between the provinces? It might be expected that industrial areas outside London stood to benefit from scientific discoveries in the capital. The evidence of knowledge transfer is frustratingly piecemeal, sometimes in ambiguous or fleeting references in correspondence; or it is circumstantial, read into individual biographies and local situations. Compared with the Scots, England was behind with its record-keeping. But another reading would be that it trailed behind in its scientific contribution to industry. Absence of evidence can indeed be evidence of absence.

---

63 Examples are given by Clow and Clow, *Chemical Revolution* – for instance, 41, 67, 136, 172, 189. The Board of Trustees for Fisheries, Manufactures and Improvements in Scotland continued its work into the twentieth century, increasingly focusing on arts and design.
64 Ashworth, *Industrial Revolution*, 145–6, 154–7. See ch. 5, 'Public Science'.
65 The Society of Arts was, from 1908, the Royal Society of Arts (RSA). It is not to be confused with the Royal Society.
66 Mortimer, *Concise Account*; Sivin, 'William Lewis', 63–4. For prizes and medals, see Khan, *Inventing Ideas*, esp. ch. 5, 'Prestige and Profit: The Royal Society of Arts'.

Engineers and scientists

In regions most associated with industrial transformation, where change accelerated towards 1800, science – in its modern definition – was still catching up. For chemistry as far as it related to ironworking, science produced more systematized explanations only after 1800. Mathematics and mechanics, while better developed as a scientific discipline, had limited application in mechanical engineering before the new century. Indeed, engineering's mechanical branch took most benefit in its early phases from those ironmaking innovations which had emerged out of industrial practice.

Before 1800 the branches of engineering were two: 'civil' engineers were so designated in order to separate them from the military, and started to gather together as a profession from the 1770s. Mechanical engineers formally gained a separate identity with their own institution in 1847. Before then, 'civil' encompassed professional engineers working in many trades. Eighteenth-century engineers were, generally speaking, less defined by specialisms.[67]

The life stories of these engineers and their professional associates are a window into how and when change occurred, and the possibilities open to individuals in their time. Individuals worked within the sequences of scientific and technological change, and a shifting industrial backdrop. This was a homogeneous group of men, privileged by education and in many cases by wealth. The group is also narrow, for their scientific and technological interests were not representative even of their own class. But collectively their experiences illuminate those times less trodden, the decades which laid foundations ahead of mass-mechanization.

These men are 'names'. In this they differ from the 'mostly unknowns' instrumental in founding a new mechanical engineering industry, which happened from the 1780s. The groups differ in other ways, too. The little-noticed machine-makers fit into that definition of 'not proper guests at the table of World-Historical Change'.[68] Their counterparts, and often their direct antecedents, had worked in forges and foundries. In contrast, the men who drove forward ironmaking and scientific endeavour, while they may have become familiar with the shop floor, in many respects lived their lives in quite different spaces.[69]

John Smeaton (1724–92), arguably the renaissance man of his age, is acknowledged as founder of the profession of civil engineering.[70] His lifetime coincided with what, with longer historical perspective, can be seen as a prelude

---

67 For definitions of 'engineer' see Cookson, *Age of Machinery*, 76–8; and below, pp. 134–5.
68 W. Ashworth, review of *Age of Machinery* in *Jnl of British Studies*, 57/4 (2018), 851–2.
69 The theme is developed further in ch. 4, below.
70 The main sources used for Smeaton's life are A.W. Skempton (ed.), *John Smeaton FRS*; Skempton, 'John Smeaton (1724–1792), civil engineer', in DNB; Denis Smith, 'John Smeaton FRS (1724–1792)', BDCE, 618–28.

to engineering's own transformation. The massed machine-makers would follow. In delivering infrastructure, and in how he freely shared key technical information, Smeaton was a generous enabler. He seized opportunities to apply his own prodigious talents to an eclectic range of innovative projects. Smeaton became the go-to figure for difficult commissions the length and breadth of the country. There was nothing, it seems, that he would not take on. His insights informed other, often overlapping, projects. Wherever Smeaton happened to be, he was closely engaged with all aspects of work in progress elsewhere, instructing his site engineers on materials and everything else. He completed the Eddystone Lighthouse, south of Plymouth, while simultaneously supervising the difficult planning and surveying of the Calder Navigation, upstream of Wakefield. Eddystone made him a master of stopping leaks, a useful talent to bring to northern canal-building; and he negotiated the sale and transfer of heavy plant from Plymouth for use on the Yorkshire job. So the ventures had significant crossover.[71]

That is only a snapshot of how any in-demand civil engineer might organize business. What sets Smeaton apart is how he accommodated science into the practice of engineering. An unusual background had liberated him from the need to conform, and his renown, established as a young man, was such that he was able to pick and choose his work. Of course, being financially self-sufficient helped.

The scope of Smeaton's interests was astonishing. He was as easy conducting experiments into waterwheel efficiency, building steam engines, producing his own drawings, or making scientific instruments as engaging in civil engineering's more routine operations: constructing bridges, harbours, lighthouses, river navigations, canals, and fen drainage systems. Two strands are of particular interest in illuminating how his own scientific research informed his technological developments. These are Smeaton's work on prime movers, and his expertise as a maker of 'philosophical instruments'. In the latter, he was partly self-taught, after declining to follow his father, a wealthy attorney in Leeds, into the law. As a successful instrument maker in London, Smeaton's reputation for brilliance grew. His professional skill allowed him to perform a remarkable series of experiments into the power of water and wind.

The occupation most associated with building and maintaining power systems and prime movers is that of the millwright, and the term embraces a great range of industrial specialisms and levels of status.[72] While some had a grounding in mathematics, millwrights largely worked empirically, using rule of thumb, the lore of the trade.

---

71 Taylor and Levon, *John Smeaton and the Calder Navigation*, esp. 23, 35, 42, 69.
72 For more on millwrights, see ch. 4.

The efficiency of different types of waterwheel had become a pressing matter. Across Europe this was contentious long before 1752, when Smeaton picked up the problem. First, he built working models, to control and calculate the variables with an accuracy impossible on site, where a constant and measurable flow of water was difficult to achieve. But scale models introduced another problem: compared with full-size waterwheels, models generated more friction, both hydraulic and mechanical. Experimental results from his apparatus could not translate directly to an actual waterwheel. Smeaton's breakthrough was mathematical, a formula allowing for variations in friction. The conclusion was a surprise even to him, and totally contradicted accepted wisdom: overshot wheels proved to be twice as efficient as undershot. Smeaton built his first watermill in Lancashire in 1753, the year that this discovery won him election as FRS. Following a number of trials to check his results, he published rules in 1759 which provided site-appropriate guidance, recommending breast wheels where site constraints did not permit overshot.[73]

This very practical solution later drew the criticism that Smeaton was insufficiently mathematical. French engineers, starting with Carnot, picked up on his findings to give their nation an early advantage in turbines. But Smeaton's methods, the fact of his experimentation, his published template for others' benefit, show him to be a most unusual engineer of his time.[74]

After these remarkable investigations on water power, Smeaton worked on windmills and steam engines. His professional base was the family home a few miles east of Leeds, though the challenging commissions took him far and wide. His fees were low. Having independent means, Smeaton chose work which interested him. His experimental findings had an early trial with a watermill built at Wakefield in 1754.[75] The design of his windmills, though few in number and concentrated near his base in Austhorpe, was widely influential.[76] Smeaton advised on the layout of the Carron ironworks and installed four watermills there, 1765–77.[77] He was also responsible for 13 steam pumping engines, mainly for collieries on the Tyne and Wear but including 1 for the Carron ironworks and 2 in Leeds in 1781, for the Middleton Colliery and Seacroft Foundry.[78]

Smeaton's home region, particularly along the Pennine range marking the Yorkshire–Lancashire border, benefited significantly from his improvements to waterwheels and water systems. Water power, far from conceding to steam, was approaching its heyday. So his findings had enormous local and regional

---

73 Smith, 'John Smeaton', DNB; Wootton, *Invention of Science*, 486–9; N. Smith, 'Scientific Work', in Skempton (ed.), *John Smeaton*, 37–41.
74 N. Smith, 'Scientific Work', in Skempton (ed.), *John Smeaton*, 48–9; D. Smith, 'John Smeaton', 620.
75 Smith, 'John Smeaton', 620, 628; N. Smith, 'Scientific Work', 37.
76 Skempton (ed.), *John Smeaton*, 77–81.
77 'Charles Gascoigne', BDCE.
78 And see Scott, 'Smeaton's Engine of 1767'.

relevance, especially for the expansion of textiles and for ironworking trades. Every one of the one hundred fulling mills operating in the West Riding in the 1770s was water-powered. From that time, dozens of water-powered cotton factories were built in the Pennines: before 1790, 61 on the Lancashire side and 41 in Yorkshire. In Sheffield, the half-century from 1725 had also seen a boom in new watermills, many driving cutlers' grinding wheels. By the 1780s all possible sites on that town's five rivers had been developed, around 136 mills in all. Later, some cutlers' workshops were replaced by water-powered forges, but the number of wheels counted in 1794 marks a high point not exceeded.[79] Waterwheel technology continued to develop because it was capable of generating more power than contemporary steam engines could, which by the mid-nineteenth century was up to 100-horse power, enough to drive a large textile factory.[80]

Smeaton also directly boosted the north's industrial potential with supportive infrastructure, notably the river navigations: he planned and built the Calder and Hebble waterways, 1757 and 1760–70, and oversaw improvements to the Aire and Calder, 1772 and 1775–9. He presented his theoretical discoveries in ways which made them easy for others to adopt, especially the site-specific guidance indicating the best-suited wheel and power system, so that his findings were instantly valuable on the ground.[81]

Smeaton is rightly seen and measured as a nationally renowned figure. While Austhorpe, close to Leeds and Wakefield, remained his home, his birthplace, and where he died aged sixty-eight in 1792, he was frequently in London, or preoccupied with faraway schemes. His local impact was just one part of the whole. But in many respects he was self-contained, not known for nurturing younger engineers, except for William Jessop, his right-hand man at Austhorpe, and his one-time protégé James Cockshutt of Wortley. Smeaton's was the world of merchants and landed gentry, and he serviced gentry estates and their need to ship coal, agricultural produce, and iron. Many of the new and refurbished watermills built using Smeaton principles were sponsored by such landowners, and the two Newcomen-style steam engines that Smeaton supplied in 1781 in his own locality were also commissioned for that gentry-led category of heavily capitalized industries.[82]

---

79 Giles and Goodall, *Yorkshire Textile Mills*, 125–32; Jenkins, *West Riding Wool Textile Industry*, 75–81; Aspin, *Water-Spinners*, 452–7, 460–3; Crossley (ed.), *Water Power on the Sheffield Rivers*, viii, xii–xv.
80 Giles and Goodall, *Yorkshire Textile Mills*, 126. See Cossons (ed.), *Rees's Manufacturing Industry*, V, 338–66, which describes advances following on from Smeaton's scientific work.
81 Smith, 'John Smeaton', 627–8. For a list of Smeaton's civil engineering works, see Skempton (ed.), *John Smeaton*, App. III.
82 Jessop took charge of the Aire and Calder improvements, 1775–9: Hadfield and Skempton, *William Jessop, Engineer*; 'William Jessop (1746–1814)', DNB.

Leeds had a circle of learned and scientific men, including (1767–73) Joseph Priestley, by then FRS on account of his electrical experiments; Priestley's immediate successor as Unitarian minister, William Wood, a founder of the Linnean Society; and the surgeon William Hey FRS, first president of the Leeds Philosophical and Literary Society in 1783, who also carried heavy responsibilities at the Leeds infirmary. These men were known to Smeaton.[83] There is, though, no known connection to the embryonic machine-making trade developing in Leeds during the last decade of his life. That new industry was based on artisans, and its members moved in quite different circles.

The philosophical society over which Hey presided lasted only three years, and not until 1819 was it reborn for the longer term.[84] But that was Leeds. In other provincial towns, there had been groups of similar type. Could these have had more obvious impacts on industrial innovation? And what about public science, the growing interest in lectures and publications among the urban middle class?[85] These are questions for a later chapter. Smeaton's own background suggests another possible route through which science could link into practicalities, via philosophical instruments. His starting point had been to reach the highest level of instrument-making, something he shared with James Watt. William Lewis, too, made some of his own experimental apparatus, an important capability for researchers in mid-eighteenth-century laboratories. Scientific instruments, including their role in lectures and demonstrations, were a key part in science's advance, though not in themselves 'science'.

Here, instrument-makers (for the purpose of discussion including makers of high-quality clocks) should be differentiated from other producers of machines and equipment. There was also a shift of emphasis within the instrument trade over a generation. William Lewis had devised, among other things, a hydrometer and furnace for experimentation. In the 1760s he shared with his readers (hoping they would emulate his scientific path) instructions on building the necessary apparatus.[86] Thirty years later, William Henry started to make equipment for sale and used his book, *Epitome of Chemistry* (1801) as a *de facto* catalogue for his wares. This was so popular that the enterprise became unmanageable and was handed over to a commercial company.[87] It appears that there was little market for such scientific apparatus in Lewis's time, but from

---

83 Cookson, *Age of Machinery*, 167, 176–7. 'Unitarian' here and below encompasses various strands of dissent.
84 Unsworth, *Leeds: Cradle of Innovation*, 110–11.
85 To be explored in ch. 5.
86 Sivin, 'William Lewis', 77.
87 Greenaway, 'William Henry', DNB. Henry was a towering figure in Manchester science: see ch. 5.

the turn of the century the instrument-making trade could and did respond to a new and substantial level of demand.[88]

Instrument-making concentrated in London throughout the period, and around the naval institutions of southern ports and dockyards, with makers present in smaller numbers in certain provincial towns. Clockmakers were universal – and habitually linked, on sparse evidence, to textile machine-making. But any role they took in textile-engineering was confined in fact to the late eighteenth century and, for instance, to producing gears for Arkwright's early experiments. The word 'clockmaker' then filled a linguistic void and was applied to makers of mechanisms for machines. The textile 'clockmaker' did not make clocks, and no textile engineer ever styled himself clockmaker.[89] There are inconsistencies to explain: Benjamin Huntsman was indeed a clockmaker, though it was on a specific quest for quality tool steel – to replace shear steel in clock springs – that he discovered crucible steel in the 1740s.[90]

John Smeaton, expert in the tools of his own trade, or trades, is best understood in the context of surveying and civil engineering, and for his theoretical accomplishments. While an exceptional figure, he was not noted for mechanical innovations. His steam pumping engines replicated technologies already in widespread use. James Watt, though – how to explain the role played by instrument-making in his quite exceptional career? Like Smeaton, Watt honed his skills in London as a youth. But Watt's father had recently failed, pressuring the son to succeed in his trade. He did that, in Glasgow, in a paid role as 'mathematical instrument maker to the university' from 1757, alongside a profitable partnership based in shop premises in the town. Watt's perspective was not that of a natural philosopher who needed instruments to advance research, though he encountered such men at the university. His work there brought him close to the professors Joseph Black and John Robison, who recognized Watt's talent as a mathematician and experimentalist. 'Everything became science in his hands,' said Robison.[91] Watt's own instruments, and some he repaired at the university, were tools which drew him into science, fascinating Watt with their evident possibilities. His good fortune, in these early academic connections and proximity to the Scottish Enlightenment, determined his later course.

In Glasgow, Watt started his steam-engine experiments, using models, from 1763. Robison had given him the idea of adapting Newcomen engines to other

---

88 See Morrison-Low, *Making Scientific Instruments*, for a definitive account of this trade in the period.
89 Cookson, *Age of Machinery*, esp. 78–85.
90 Brearley, *Steel-Makers*, 5–10; Evans and Rydén describe the difficulties with shear (or Newcastle/Crowley's) steel, in tool- and file-making: *Baltic Iron*, 140–3, 149.
91 Tann, 'James Watt (1736–1819)', DNB. Watt's grandfather had been a professor of mathematics: Marsden, *Watt's Perfect Engine*, 9.

applications in *c.* 1759. Watt had already concluded that Desaguliers, whose *Course of Experimental Philosophy* (1734–44) was considered definitive on the subject of steam engines, had made mistakes. Watt joined with Dr John Roebuck, a chemist whose industrial interests included a stake in the Carron ironworks. It was Roebuck who funded the patent of Watt's separate condenser in 1769 and, in overstretching himself and becoming insolvent, prompted Watt's move to Birmingham after Matthew Boulton acquired Roebuck's two-thirds share in the patent.

But before this, Joseph Black, the 'pneumatic chemist' interested in gases and heat, had set Watt on the path to a breakthrough by suggesting that 'economy of steam' was most important in the new design. Black, who may have introduced Roebuck to Watt, had independently identified latent heat. Theory here came together with technological design, and there emerged an engine powered wholly by steam rather than part steam, part atmospheric pressure. Watt later downplayed the contributions Robison and Black had made to his innovations. Beyond any doubt, borrowings and influences had run back and forth over the years of experimentation. By the time Watt's two famous breakthroughs entered commercial production, they had drawn on older technologies and theories, newly minted elements of theory and design, and numbers of financiers, scientists, facilitators, and other connections. All of these framed Watt's own skills and understanding, and enabled his accomplishment.[92]

Nor was Watt's proficiency constrained by the walls of a laboratory or instrument workshop, for it extended into civil engineering and chemistry. In his Glasgow years he was employed at times on surveys for canal construction, adapting and devising instrumentation for the purpose.[93] Later, in the 1780s, he developed a close interest in chlorine, newly identified, and the possibilities of using it as a bleaching agent for Scottish linens. This prompted a trip to Paris in 1786 to meet the man who had made chlorine's properties known, Berthollet, with whom he afterwards corresponded in French. In 1788, Watt carried out a trial on 1,500 yards of linen on his father-in-law's bleaching fields in Glasgow.[94]

Naturally, Watt will forever be defined by his steam engines. As we see, these had derived from a heady mix of knowledge and expertise gleaned from numerous individuals, and involving many specialisms – commercial as well as industrial and scientific. Watt was extremely well connected, and further consolidated his already impressive circle of acquaintance in Birmingham from 1774.

---

92   Marsden, *Watt's Perfect Engine*, 19–21, 31, 46, ch. 3, and *passim*; Ashworth, *Industrial Revolution*, 175–7; Campbell, 'John Roebuck (bap. 1718–d. 1794)', DNB; BP 913. For Black, see ch. 2.
93   Tann, 'James Watt', DNB.
94   Clow and Clow, *Chemical Revolution*, 186; Mantoux, *Industrial Revolution*, 250–1, fn.; Ashworth, *Industrial Revolution*, 163.

One sphere of his operation continued to frustrate Watt. Workmen capable of building steam engines were in short supply. Boulton & Watt struggled to recruit and retain men who could do what was required. This problem was widespread among makers of early steam engines. For Newcomen-style pumping engines, which existed for decades before Watt, there was a pool of apparently proficient makers, especially in mining districts. But the fact is that all steam engines were difficult to build, and even more difficult to make work. Ironically, the problem lay in precision, and even Watt, as an instrument maker accustomed to high degrees of accuracy, could not fully fix that. Counter-intuitively, achieving acceptable tolerances was actually proving more difficult on this large scale than in instrument-making.

The Boulton & Watt company fine-tuned its systems to achieve the best result that was practically possible. The quality of components was improved by a division of labour, concentrating individual workers on specific tasks. Yet, however this was cut, the truth was that the process still rested largely on handicraft skills. The usual deal was that the company supplied advice and plans, certain parts, and engine erectors, while customers sourced boilers and pistons locally, if that was possible. The main input of expertise came through cylinders bored by their contractor John Wilkinson. This was during Boulton & Watt's early years in business, before a tempestuous finale when Wilkinson was discovered to have pirated their engines. Still this problem persisted: that once all components were assembled on site, there came a period of adjustment, sometimes days and nights for weeks on end, until the engine functioned. Commissioning a steam engine involved intuition and commitment by the engine erectors who nursed it into life.[95]

And the engines, once in operation, were far from perfect. Watt's concept was let down by the inadequacy of materials then available. As he himself said, when argument raged over the originality of Arkwright's frame, 'whoever invented spinning Arkwright certainly had the merit of performing the most difficult part, which was the making of it usefull'.[96] This, written by Watt in 1784, encapsulates his own recent exertions to improve the performance and efficiency of his steam engine.

To understand more of what was happening, move on to Boulton & Watt's well-documented and unpleasant spat with Matthew Murray around the turn of the century. The real state of affairs sometimes emerges out of what came

---

95 Cookson, *Age of Machinery*, 133–4, 138–9; Marsden, *Watt's Perfect Engine*, esp. 40; Ashworth, *Industrial Revolution*, 175–6.
96 University of Glasgow, Archives & Special Coll., GB 247 MS Gen 501/27, letter 30 Oct. 1784, James Watt to his father-in-law James McGrigor, Glasgow. The idea of roller spinning came to Arkwright from Thomas Highs of Leigh, or via Highs' assistant John Kay. Watt was no admirer of Arkwright – 'he is to say no worse one of the most self sufficient ignorant men I have ever met with' – but acknowledged what the nation owed to him.

next: another party takes forward the technology, revealing problems inherent in previous versions. So it was with Murray, who represented a leap forward in steam-engine technology.

Some challenge to Boulton & Watt's supremacy might have emerged sooner, had they not so vigorously protected their patents. Like Arkwright's, their business strategy was to license products and chase down pirates, while collecting healthy profits from the hardware. Once the company's patents approached expiry in 1800, this policy proved less sustainable. Ruses to circumvent them were employed by other engine-makers, including Murray, whose trick was to supply condensers separately from engines, thus transferring the risk of legal challenge to the purchaser.

The pace of innovation at the company's Soho Foundry in Smethwick, west Midlands, had slowed, the old men gradually handing control to their sons, Matthew Boulton jun. and James Watt jun. Murray pulled ahead of the field, patenting several novel features, 1799–1802. He eliminated carpentry from steam engines, building them almost entirely from new varieties of cast and wrought iron. His firm's advantage was rooted in excellent production techniques, especially its celebrated foundry work and the quality of its greensand casting. Soon after starting to make steam engines in 1797, Murray came into conflict with the Birmingham partners. Boulton & Watt, still using timber framing, and recognizing their own deficiencies in casting, took this battle to the foundry floor. They tried to poach foundrymen and gather technical information from Murray, and later went to court, successfully overturning Murray's patents. After a bruising, expensive, and distracting encounter, Murray stopped using patents.[97] But he had made his point and asserted his superiority as an engineer. The younger generation of Boulton & Watt turned their attention to new ventures, among those gas engineering.

What to make of all this? For steam engines – and for textile machinery and machine tools – transformations in ironmaking in and around the 1790s marked a defining moment. The impossible would become possible. Technologies could be accomplished which had not previously been practical options. In heavier equipment, cast iron replaced carpentry. The strength and rigidity of all-iron construction were as important as the precision which it enabled. Carpenters were sidelined in this respect, and the component-makers' workshops took centre stage. By the 1820s, many of these businesses had transformed into machine-making factories with foundries. There, Matthew Murray was a decade or more ahead, gaining further advantage by working in both wrought- and cast-iron production, and on his own-make machine tools. Murray's organizational changes, and those of others who followed, would not have

---

97 Cookson, *Age of Machinery*, 159, 205–8; also 226–30, 'Making things work'. Marsden, *Watt's Perfect Engine*, describes the long process of converting Watt's design into a workable reality.

been possible without machine tools. Nor would the finer textile machines mass-produced from the 1820s, the self-acting mule, power loom, or, later, mechanized wool-combing.[98]

Matthew Murray was an outstanding engineer of his time, a liminal point for machine-making. Mechanical engineering stood on the threshold of branching into a trade in itself.[99] In this, Murray's contribution was significant. Boosted by recent advances in ironmaking, he improved his own flax-spinning machines and machine tools, carried steam engines into a fresh era, and pioneered steam-powered railway systems. He also laid an early template for the engineering factory. Murray built from two elements: new concepts, and skill in production. 'Skill' involved technique and technical knowledge, and increasingly the novelty of machine making machine. Not every process could then be mechanized. So in the design and making of new forms of machinery, conceptual and manual skill, though evolving, continued to be of the utmost importance.

The career of Murray, by trade a whitesmith, is a fine representation of this, built on a mix of inspiration and artisan proficiency. A similar dynamic played out in iron production. It was all about 'the making of it usefull'. 'Things going wrong' was also a necessary part of the exercise. The mid to late eighteenth century was an era of false starts in science and technology. Failure and inadequacy are not a by-product of innovation, but fundamental to it. Without things going wrong, how to see what is right? What works? What is the best outcome? The learning comes from mistakes, our own and others'.

Just as chemistry was not quite in tune with iron production, mathematical theory did not entirely harmonize with machine-making. It must have seemed that chemistry, its fundamentals then underdetermined, only partially understood, had less tangible value to an iron forge than its intuitive smiths and the blue flame. The problem for mechanical workshops was different: theory was better developed and might have made the smiths' job simpler, had they only known about it. So neither science nor technology ran fully ahead of the other. Chemists could, like Lewis, advise and help work out solutions in iron production. Smeaton used mathematics to resolve mechanical questions, brilliantly so, but was rooted in older forms of technology and not engaged with the new machinery emerging in his final years. Meanwhile certain 'scientifically-minded industrialists', among them Joseph Dawson of Low Moor, James Watt and David Mushet – all products, one way or another, of Glasgow – did connect

---

98 I am indebted here to Ron Fitzgerald, whose review of *The Age of Machinery* included significant ideas and information on this subject: *Int. Jnl. for the History of Engineering and Technology* (2019), 1–4. https://doi.org/10.1080/17581206.2019.1575086.
99 The Institution of Mechanical Engineers was founded in Birmingham in 1847, when the industry was already large and the profession well recognized.

with academic science and philosophical societies.[100] The nature of those links will be further explored.

Advances in machine-making and iron production achieved before 1800 owed much to local ecologies of innovation. Highly significant were skills honed in practice, the 'blue flame' knowledge, manual dexterity, strength, and judgement. Artisan skill absorbed new forms of expertise. But skill alone would not solve all the practical problems. Watt struggled to make his steam engines work efficiently: timber and leather were inadequate materials for the task, while the associated science, the underlying principles of steam engineering, were only just then developing alongside the technology. Practice informed science as much as it rested on it.[101]

---

100 Pacey, 'Emerging from the Museum', esp. 458. The phrase comes from Musson and Robinson, *Science and Technology*, ch. III.
101 Ashworth, *Industrial Revolution*, 175–6.

4

Gentlemen and players

Searching the archives of heavy industry can be a dirty business. In this, the Middleton Colliery records do not disappoint. Their accounts, showing purchases of iron products and sales of coal to Hunslet Foundry in the 1770s, are filthy. Partly this is an outcome of age, but mainly it is coal dust, inescapable even in the clerk's office, where the neatly kept books were in daily use.

This triggered an idea, in the way that connecting with raw archives can do, an inkling about the document's creation. The dirt sparked memories of weeks spent working on Richard Hattersley's similarly grimy business papers in Bradford archives. Yet there are striking differences between the Middleton books, the work of a trained and methodical scribe, and the much less orderly entries, in his own untidy writing and phonetic spelling, of the whitesmith Hattersley. At the time, starting in 1793 when Hattersley stepped up to managing partner, his business was becoming hub of Keighley's embryonic machine-making trade, which would later achieve world renown.[1]

Here is more than a snapshot of two contrasting workplaces. These industries stood in proximity – in time, location, and technology – and their mutual links were growing. They had grime in common, but what else? How deeply did the similarities extend? Coal and iron, on the one hand, and machine-making, on the other, had emerged through contrasting pathways. Local contexts influenced work choices, and determined the shape and capabilities of key industries. But the choices available to individuals, where they fitted into those businesses, were constrained by where they stood in society. Making sense of events, then, requires some explanation of social intricacies. Those human relationships were as significant to what took place as were technologies and economics.

In the main, Hattersley and other newly minted engineers setting up in business from the 1790s came out of artisanal metal trades. They were intelligent and skilled, although not highly educated, and their firms built from small beginnings on reinvested profits and modest finance from partners.[2] In

---

1 WYAS Leeds, WYL 899/222, Middleton Colliery ledger 1780–1; WYAS Bradford, 32D83/5/1, Richard Hattersley's day book, 1793–8. Middleton was three miles from the town of Leeds.
2 Cookson, *Age of Machinery*, esp. discussion on pp. 108–27.

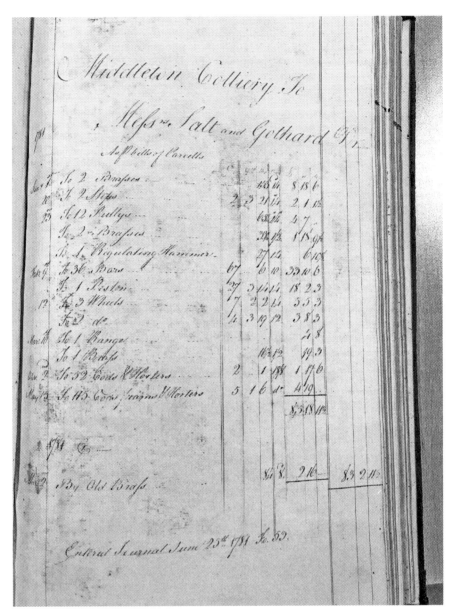

Plate 7. The Middleton Colliery ledger, here recording transactions with Salt & Gothard's Hunslet Foundry in 1781, shows a well-educated clerk working in the dirty conditions of a colliery office. Source: WYAS Leeds, WYL 899/222, extract from Middleton Colliery ledger 1780–1.

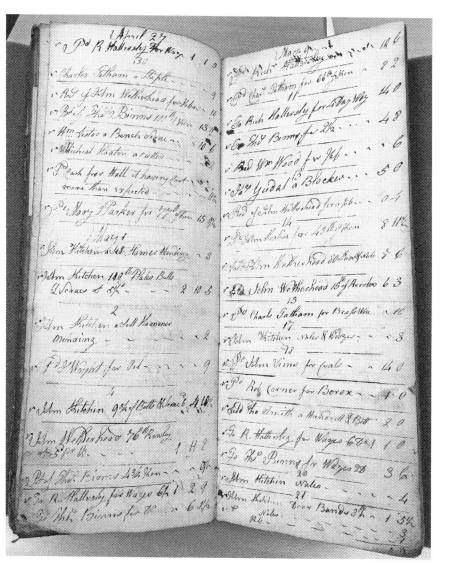

Plate 8. In marked contrast with the Middleton clerk, Richard Hattersley's records – here, a sample from when he first took over the Keighley business in 1793 – suggest a man numerate and literate, but educated only as far as considered necessary to his rank. Source: WYAS Bradford 32D83/5/1, extract from Richard Hattersley daybook, 1793–8.

the industries with which they had dealings, though, traditions and influences differed. The urban textile scene and the iron trade, discussed in earlier chapters, themselves made weighty contributions to a bigger technological and economic picture. So too did coal-mining. But in that case, and other extractive enterprises, the ultimate proprietors were often wealthy landowners at arm's length, the businesses largely managed by agents. Between them and the leaseholders, a pool of specialist intermediaries found employment: managers, clerks, technical advisers, lawyers, and other professionals. Some of that group, loosely categorized here as 'intermediate', while neither leading industrialist nor artisan, had considerable influence, some even emerging as catalysts of industrial change. So they are of particular interest.

Who comprised this quite diverse 'intermediate' group, what exactly did they do, and how? 'Intermediate' is disparate enough to encompass the Middleton clerk and the colliery's managers and viewers, millwrights certainly, and iron producers of a new stamp such as the Walkers of Rotherham, and Salt & Gothard in Hunslet. These categories represent a range of skills and training; some of the individuals had access to funds, some were consultants or subcontractors, and others were salaried. Some had clean hands, while others became dirty in the course of working, and indeed it was possible to be both grimy and wealthy. Altogether they administered, supervised, and innovated on behalf of colliery and ironworks owners. Some established spin-off businesses based on what they had learned elsewhere. They might be involved in supplying materials or parts, or in offering technical and other support to early factory projectors, and to the machine-makers who brought a new focus to metal-working. They were enablers or instigators of actions and transactions fundamental to industrial advance. And in 'advance' can be understood anything surrounding industry, within its supportive context. That might stretch beyond technology and transport and buildings, to financial and administrative systems and more.

While trade or profession is some marker of social status, it is far from definitive, especially in an era of great industrial change. So the progress of certain occupations will be discussed – paying particular attention to the millwright, for reasons to be explained. Social networks, so profoundly important in how matters worked out at the time, are a strong indication of position within the hierarchy. The British class system is mystifying, and not only to foreigners. Even now, it has considerable traction in determining life chances, though as a day-to-day concern it may feel less immediate and less finely graded than it has in the past. The eighteenth-century person, though, was finely attuned to rank and manners, this tacit understanding of their place in society instilled early on. To the modern-day investigator, individuals' situations in past societies are less instinctively obvious. But keeping in view Thompson's point that class is a spectrum rather than a separation, the clues are there, especially in social networks, the company a person kept. Here are routes into finding more about the nature of work and its relationships, and so about industrial innovation. The answers rest in specifics, real people in real circumstances.

This brings to mind E.P. Thompson's famous 'enormous condescension of posterity'.[3] He meant that overlooked groups, routinely dismissed from a bigger historical picture, should be rescued from such disdain. Thompson was not being kind or sentimental, nor was he trying to create new heroes. His point was about finding the truth: it is incumbent on historians to seek clarity, to see what people actually did and how they related to each other.

The condescending comes not only from being ignored or sidelined. Accomplishments of people in history are often diminished and misrepresented by lazy terms used to describe them. Labels can confuse and offend. Is there any need for 'heroes', or 'tinkerers' (even 'inspired' ones)? Landes may have been first to claim that Britain's strength was built on 'practical tinkerers'.[4] 'Tinker' was originally an acceptable alternative name for a tinsmith, but 'tinkerer' now has overtones of messy amateurism which would affront a skilled worker. More than that, without further explanation it is misleading. It diminishes skilled people who were doing something notable in different and difficult times. But its greatest offence is that it does not advance an understanding of what was actually happening.

So how did these relationships work in Britain's key industries? How did sectors connect? 'One size fits all' is not to be assumed – that all regions, industries, trades, progressed similarly. The reality of industrial life is the theme here, a search for the forces at work.

Class and its consequence

Social class determined access to occupation, education, and useful networks. While the discussion which follows concentrates largely on northern England, its findings are more widely applicable. Localities varied in their industries and working arrangements; in social status terms, though, the same broad categories hold true across Britain. And there was fluidity: over time, an occupation might lose or gain prestige, new industrial groups appear, or individuals occasionally move within the hierarchy, perhaps through marriage or growing affluence.

It was therefore complicated, and not simply defined by trade or profession. By custom, though, that is how a working man would announce himself, retaining that original label even when he settled into something quite new. Richard Hattersley and Matthew Murray continued to be known as smith, or whitesmith, long after they had become machine-makers. Apprenticeship

---

3 Thompson, *Making of the English Working Class*, 13.
4 Coleman, 'Textile Growth', 13.

conferred rights to assistance under the Poor Law in the parish where it had been served. This was important to individuals, and potentially a lifeline for a family in distress. It also guaranteed that a migrant would not become a liability to their new parish. The tradition of outdated self-identification endured far into the nineteenth century, and the self-descriptions should not be taken at face value.

But also 'smith' can mean many things. So too, said Thompson, can the label that covered dozens of trades, 'artisan'. This embraced everything, from a prosperous master craftsman with a large degree of independence and many employees to sweated garret labourers without autonomy. Within one industry – Thompson cites carriage-making, but examples abound – could be found many grades of skill. And of course the stature of various apprenticeships, and of whole industries, shifted over the decades. Thompson sees an old élite, of 'master-artisans who considered themselves as "good" as masters, shop-keepers, or professional men', and a new élite whose expertise in iron, engineering, and manufacturing boosted their previous relative position in the pecking order.[5]

This 'middling' set, along with the 'intermediates' described above, must be brought into focus, especially for the period leading up to about 1780. Best avoided is the term 'middle class', a confusing anachronism stretched beyond any useful meaning. In contexts of unprecedented change, the positions of these disparate actors were not static. What applied in one generation might shift markedly for the next. So what follows is an endeavour to see people for what they were, and set them in their own time.[6]

Why are the standing and competences of these groups so important? It matters, because in the eighteenth century it mattered. Nuance and connection counted for a lot, smoothing the way in social and professional life. Aristocracy and gentry are recognizable by titles and ties. In some cases the professional class – servants of the better-off, like agents and lawyers, and also high-status merchants, clergymen, and physicians – were themselves gentry. Whether they were or not, they were entitled 'Mr'. But so too, called 'Mr' and 'yeoman', were consultants such as master millwrights, prosperous business owners (millers perhaps, and high-end retailers), and propertied farmers. Then came 'small masters', like those of Sheffield, a workshop-based, apprentice-trained class with in-demand skills and some resource and independence. After that there were employees and subcontractors, and then workers classed as having lower skills. Agricultural labourers counted there, though the idea that they were all unskilled is a travesty. Then came the very poor, scrambling opportunistically

---

5 Thompson, *Making of the English Working Class*, ch. 8, esp. 259–63.
6 This marks a fundamental difference in approach from Mokyr, *Enlightened Economy*, 2, 9–11, 14–17, particularly in how society was structured, what people knew, and how industries and science were related.

for a living in any way they could find. But none of this was easily enumerated or expressed in terms of income, as even the resolute contemporary statistical researcher Joseph Massie would find.[7]

For every one of the assertions above, exceptions will be found – variations across British regions, and between rural and urban settings, guild and non-guild towns. There are, however, some broad truths across the 'middling' set, the Mr and the yeoman. Their education and training, probably with a commercial and economic component, was superior to that of the general run of workshop-based mechanics. Any profession – because usually that is what it was, a clear notch higher than an artisan trade – would be acquired through a specific path, often a premium apprenticeship, for which a fee (sometimes considerable) was levied.

Even within the ranks of those professionally trained, there were obvious gradations. One already encountered is that of the Elams, up-and-coming export traders of cloth, as they related to the grandest of Leeds woollen-cloth merchants.[8] Similarly, Hunslet Foundry's millwright-turned-ironfounder of 1770 dealt with collieries whose owners (though inconspicuous day to day) stood on the cusp of aristocracy and higher gentry. The colliery agent, the *de facto* manager, while close in background to the millwright, was probably considered higher-ranking because he represented people of status.[9]

Status, as any reader of Jane Austen knows, did not correlate exactly to wealth. A gentleman's daughter, while disadvantaged in her prospects if she lacked money, was still considered a suitable connection, even as a bride. An impoverished governess would dine with the family, not the servants. Class was apparent in a broad sense from the way people dressed, spoke, carried themselves. Memoirs and novels over the longer period of industrialization show this, and further confirm how such relationships shifted over time.[10] Roberts describes 'the English proletarian caste system in all its late flower' during his childhood in Edwardian Salford.[11] Class within class endured, even in the lower strata.

The likes of the Elams did not share the social status of the older merchant-house principals in Leeds. Such men, says Wilson, were 'seldom acceptable in the corporation or in the well-established apprenticed merchants' dining rooms'.[12] Ironically, even the immensely wealthy gentlemen merchants from

---

7 Mathias, 'Social Structure in the Eighteenth Century'.
8 Wilson, *Gentlemen Merchants*. See also above, pp. 19–20.
9 Cookson, 'Hunslet Foundry'.
10 *Emma*, or any other of Austen's works, is a guide to social niceties, *c.* 1800. So (though later) are the Brontë novels, especially on governesses (see Charlotte Brontë's *Jane Eyre*); and George Eliot's *Middlemarch*, a subtle treatment of social status.
11 Roberts, *Classic Slum*, ch. 1.
12 Wilson, *Gentlemen Merchants*, 32–3.

the older Leeds families were to a degree looked down upon as 'trade' by Yorkshire's landowning gentry, some of whom were far less financially robust. Tensions between commerce and manufacture, land-based and urban wealth, were at the root of political turbulence and electoral reform in the nineteenth century. By that time the previous model of Leeds merchant was defunct, and the individuals had themselves decamped to country estates.[13]

So some fluidity in social standing is evident, quickening as 1800 approached. Industrial hierarchies shifted to accommodate new ventures and new kinds of occupation. That includes, in the mid-eighteenth century, a different breed of innovative ironmaster emerging from secondary metal industries in Yorkshire and other regions. These initiatives were relatively few, but highly influential in the broader economy. It was still possible then to build a business from very little, as illustrated by the Walkers of Grenoside and then Rotherham, discussed in chapter 2. A generation later, such self-funded ventures were far beyond the capacity, financial or technical, of any unsupported individual.

While the barriers into business were lower for smiths and others entering mechanical engineering, these men must adapt significantly before they became acceptable to polite society. Many doors remained closed to those with rough manners, clothes, and speech. Even the pioneering railway engineer George Stephenson found his origins a source of embarrassment, and an impediment throughout his life. Trained as a colliery engineman, his parents illiterate, he taught himself to read and write in his late teens. But he was never good at writing, and despite some tuition in arithmetic did not master theoretical calculation. Stephenson was very conscious of the social disadvantage of a strong Northumbrian accent, and in consequence declined many invitations and honours. He ensured a better education for his son, and as a young man the refined Robert undertook the company's public-facing presentations, such as appearing before parliamentary committees.[14]

This was a role of the intermediary, the agent, manager, or professional who facilitated communications across levels of hierarchy, often while promoting their own schemes and initiatives. Through these routes, it was possible to gain access to social circles previously denied. A confident and necessary man, someone like the Leeds master millwright John Jubb, at ease on the workshop floor, on site overseeing earthworks for a factory, and also in a more elevated world closed to artisans, might progress socially as his business, reputation and fortunes flourished.

In the turmoil of endemic change, there were ebbs and flows too in the stock of certain occupations. 'Class' remained firmly entrenched, in terms of a structure which bestowed and also limited opportunities. It was necessary,

---

13 See above, pp. 17–20.
14 Kirby, 'George Stephenson (1781–1848)', DNB; 'Robert Stephenson (1803–59)', DNB.

though, that the hierarchy adjusted to accommodate novel forms of work in the new, rising industries. Industrialization influenced social relations, but the path which industrialization had taken was already heavily determined by the social class of the groups at work within it. The reason is that access to education, scientific ideas, and practical training depended upon social rank and family resources. Class generally trumped cash, and often ways were devised to place the sons of higher-status families into education and professional training even if personal funds were short. Status was almost entirely fixed by birth, though subject to some amendment – by marriage, perhaps, or the prestige of professional training. But none of this was easy, and connections were everything.

At certain points, artisan businesses gained substantial agency in developing key technologies and systems. Thus it was conceivable for the educated but relatively humble (by birth if not in ambition) Walkers to become innovative ironmaking magnates, and forty years after that for a machine-maker like Murray to transform his own situation. This happened not only because he was Matthew Murray, and a shining light among mechanical engineers. There were others of his type, if not matching his technical brilliance, in Murray's neighbourhood and in comparable northern industrial towns. The main assets these men possessed were metal-working skill and mechanical aptitude. They also benefited from opportunity, a ready market for their products, and that measure of agency. Above all, they had the advantage of time and place, their position in a workshop network. They were part of a group, set in an adaptable and innovation-focused locality and industry.[15]

Demonstrably, many of the more significant individuals in textile-engineering's formative networks were neither much educated, nor financially well resourced. But evidence points to broadening horizons, an enlightenment within workshop communities in the late eighteenth century. The spirit of enquiry was alive. Older trade links into northern Europe – through Swedish and Russian iron imports, and coal exporting to the Baltic and Netherlands – were explored by the new engineers, with Murray again in the lead. Curiosity about the world, and beyond, was not only a matter of business. The inner lives of machine-makers and their neighbours extended far and wide. They surveyed creation with serious intent, through ideas that were religious, political, philosophical. The towns bore witness to a range of new creeds, of Methodist factions, Unitarian and Swedenborgian congregations, Freemasons' lodges.[16] This was the ecology of mechanical innovation.

Poor literacy did not reflect a lack of intellect, though it was a great impediment to anyone hoping to access scientific literature and works in other languages. The reality for most was access only to the basics in a local school,

---

15 See apps 1 and 2 in Cookson, *Age of Machinery*; also the Sheffield workshops model in ch. 6, below.
16 Cookson, *Age of Machinery*, 169–73.

perhaps for a year or two if their family had the means. And that, at best, was the situation in Britain before Forster's Elementary Education Act in 1870. In general, more adults could read than write, and if they were equipped to keep simple records, it was probably not with ease and fluency.[17]

'Petty masters', as Thompson calls them, needed to write in order to keep business accounts. Any artisan in these industries – whether smith, founder, joiner, or other skilled manual worker – must at least read and work with figures. Because many firms were tiny or transient, holding on to records in the longer term could not have been a priority. Even with better-known textile engineers, business records of the early period are often lost, or, like Murray's, destroyed in a later factory fire or other mishap. The survivals are physical affirmation of the diversity and competences of people at work there.

Returning to the contrast between a Middleton Colliery clerk and the rough and ready writing and spelling of one of Keighley's foremost textile engineers: Richard Hattersley, a smith from a nail-making background, afterwards a striker in a Sheffield forge, arrived at Kirkstall to manage the new screw forge there in 1788. His had become a skillset in high demand, and it seems that he was quickly poached by a Kirkstall Forge customer who needed him in Keighley.[18] Hattersley was typical of small masters in his trade: shrewd, independent, technically proficient, but with limited education and seemingly little need of (or access to) published works. In all this, he was characteristic of the artisan class of Sheffield, Keighley, and elsewhere.

The unnamed clerk at Middleton typifies something else about those highly capitalized industries, mining and ironworking. While closely associated with early mechanical engineering, they organized themselves quite differently. The larger ventures accommodated – in itself evidence that these were needed – that intermediate layer of managers, technical advisers, and record-keepers. The owner's representative – the agent, colliery viewer, manager, or perhaps a combination of those roles – supervised other middle-ranking employees including the educated clerks responsible for record-keeping. In ironworks starting up with fewer resources, like the Walkers in the 1740s, Hunslet's Cupola Company in 1770, and the reborn Kirkstall Forge under Butler & Beecroft from 1779, it was more likely that owners, family members, or a partner would initially shoulder the burden of management and technological input, at least until it was possible to pay for assistance.[19]

As for the new mechanical engineering, this was still, before about 1815, some distance away from adopting factory-based mass production as the norm. Its surrounding community, a supportive context of suppliers and customers,

17 Thompson, *Making of the English Working Class*, ch. 16, esp. 783, 787.
18 Cookson, *Age of Machinery*, 91–4, 165.
19 Butler, *History of Kirkstall Forge*; Butler, Butler, and Butler (eds), *Diary of Thomas Butler of Kirkstall*; Cookson, 'Hunslet Foundry'; also ch. 2, above, 'Building resilience'.

subcontractors, credit, knowledge, and information, remained important, and not only in commercial and technological ways. The religious fluidity on display, minds that opened to new enquiries and concepts, discovering what was fixed and what negotiable – this age of inquiry carried such infant industries onwards. Whichever religious domination or political view an individual might settle upon was rather less significant than the fact that such speculations were actually taking place.

Before addressing the extent to which enlightenment science penetrated this world – that is, the reach of 'public' science and its influence on key industries – there is another matter to settle. Confusion about millwrights, where they stood in the industrial and social order, and their impact on industrialization, is long-standing. As they are the source of persisting mythology, millwrights need to be better understood.

Explaining millwrights

In any account of industrialization, millwrights are likely to take some significant role. Yet they are routinely misinterpreted.[20] What precisely was it that millwrights did, before, during, and after the upheavals in the industries that they served? Millwrights came out of skill traditions that were markedly different from those of machine-makers (in the modern, post-1790, iron-framed sense of machinery and machine tools). They continued as distinct and separate trades. The story of millwrighting is also a lesson about social class in industry.

There is complexity in millwrighting, a centuries-old occupation widespread across Britain. The work changed over time, and its substance differed from place to place. Much was site-based, and millwrights tended to specialize in the industries and power systems prevailing in their own locality. Grinding food stuffs was common in mills everywhere, while certain regions produced specialist millwrights dedicated purely to ironworks or mines, or working largely in the fulling and frizzing mills of woollen textiles. As a trade, millwrights were renowned for adaptability, for finding solutions as industry changed. The eighteenth century saw some drawn into particularly dynamic situations. There is no uncomplicated way to define 'millwright' without reference to time and circumstance.[21]

A source of historians' confusion lies in the 'millwrights' work' mentioned in textile factory insurance policies. From the 1770s, millwrights took a lead

---

20 The tradition of misconstruing millwrights continues: Mokyr, Sarid and van der Beek, 'Wheels of Change', has been published since this section was written.

21 For instance, the OED gives 'millwright' a current, modern, sense, one who installs and repairs machinery. This may be so, but was not the reality of the eighteenth-century millwright.

role in designing, building and maintaining factories, waterwheels, dams, and water courses. They installed steam engines, and gear wheels and shafting for transmission systems. The old style of millwright was also responsible for a mill's manufacturing apparatus, often millstones, though in fulling mills it was the heavy wooden stocks used to pound woollen cloth during finishing.

As textile processes gradually mechanized, a new demarcation was drawn within the mill or factory. Over this boundary, where transmission met machinery, the millwright gave way to a finer kind of mechanism. This, insurers first called 'clockmakers' work', though only most exceptionally did it involve clockmakers. While the term stuck, the clockmakers (if they had ever engaged) did not. Here was the earliest textile machinery, made using precision components – rollers, spindles, flyers – produced individually by smiths. They worked to tolerances so fine as to be unattainable by millwrights. These first machines were framed in wood, the frames sometimes produced by millwrights and assembled into machinery with precision parts supplied by subcontractors. But from the 1790s textile machinery began to be powered and iron-framed, and was produced by increasingly specialized metalworkers, often the same smiths who had first developed fine metal components. This marks out smiths as the foremost initiators of what we now call mechanical engineering.

Before 1800, the suffix 'wright' meant woodworker, while 'smith' (blacksmith, tinsmith, whitesmith) was a metalworker. To say that millwrights were carpenters is not to diminish them. But the division in traditions is clear. Millwrights, or milnewrights, were as old as the mills, the trade found in Yorkshire sources from 1387 and recognized by Act of Parliament in 1563.[22] With millwrights, the emphasis is on 'wright' – other examples are ship-wright, cart-wright, wheel-wright – indicating a carpentry-based trade.[23] 'Engineer' was little used except in reference to the military, until 1771, when the Smeatonian Society started to talk of civil engineering. Then, from about the 1790s, 'engineer' came to indicate a maker of engines or machinery.[24]

This background challenges the idea that millwrights were 'mechanics', in the sense of mechanic as a worker in metal.[25] Windmill-like cranes to be seen in Hanseatic ports show what early-modern millwrights could construct,

---

22 OED; *Yorkshire Historical Dictionary*: http://yorkshiredictionary.york.ac.uk/; Moher, 'London millwrights', 6; Cookson, *Age of Machinery*, 71.
23 OED. The OED does not link 'wrought' with 'wright', though wrought means worked, hence wrought iron. The *Yorkshire Historical Dictionary* has usages of 'wrought' from the fifteenth century, meaning worked by hand, or worked up; and 'unwrought' meaning pits or coals unworked.
24 Cookson, *Age of Machinery*, 76–8. Pole has much more to say about the long history of 'engineer': *Life of Sir William Fairbairn*, 3–7.
25 See, for instance, Mokyr, Sarid and van der Beek, 'Wheels of Change', which does not address the key point of what 'mechanic' or 'mechanical' actually meant.

applying tried and tested techniques to new formats.[26] The switch made by many millwrights – to use iron, particularly cast iron for structural applications, supplied by specialist foundries – was demonstrably a phenomenon of the late eighteenth century and after. John Rennie's novel endeavour to introduce cast-iron components into millwork, from the mid-1780s, is well documented. During three months spent with Boulton & Watt in 1784, Rennie investigated sources and potential uses for such parts, his plans based on calculations of the specific power needs of certain processes. It was a systematic mission related to constructing the pioneering Albion Mill in Southwark, London. Boulton & Watt's interest was in demonstrating their rotative engine's potential for flour-milling. In his own field, Rennie made important breakthroughs at Albion Mill. He substituted cast-iron pinions for wooden ones in the mill gearing, based on mathematical calculation. Wooden cogs were not eliminated, and the design was not perfect. But William Fairbairn, who later picked up Rennie's baton on millwork, saw this as 'the greatest advance in the application of gearing', enabling far larger and more efficient waterwheels to be constructed.[27] Here is a snapshot of the considerations then occupying progressive millwrights.

The ensuing, nineteenth-century, interpretations of millwrighting have coloured and obscured the view, though there were eighteenth-century inconsistencies too. William Fairbairn's musings on the subject, written in the 1860s, are still much quoted. Historians have fallen gratefully upon scraps, and presented a variety of interpretations of millwrights' work and abilities. There are significant discrepancies to be revisited and explained. One concerns the problem of terminology with the word 'mechanic', which to the modern ear evokes 'metal'. But how did the eighteenth century understand 'mechanical'?

First, though, if it is contended that the millwright was a trade apart from textile-engineering, how do we account for the involvement of millwrights in machine-making? There is no doubt that some millwrights – only a few, but in circumstances which must be explained – constructed textile machinery before 1800. Where this happened, the frames were wooden, and the precision parts were made by smiths, whether directly employed by the millwright or machine-maker, or subcontracted to specialist component-makers. In the case of John Jubb (1748–1808) of Leeds, a wide-ranging millwright whose textile interests started with a share in an early cotton factory, all of these applied: he bought machine parts, spindles, and screws from Kirkstall Forge (recorded from 1787); sold basic textile machinery, scribbling and carding machines, and willeys, and also the millwright product, fulling stocks; employed millwrights, machine-makers, and joiners and whitesmiths with experience of textile machinery;

---

26 For example, the reconstructions in Stade, Lower Saxony (based on a crane dated 1661), and in Rostock on the Baltic coast.
27 Moher, 'London millwrights', 153–69. The father of William Murdock played some part in introducing cast-iron parts to mill gearing: see below, p. 126.

and trained apprentices, both premium (millwrights) and trade (mechanics and other artisan trades).[28] After a series of early deaths, the Jubb firm folded. Might Jubb's machine-making have continued? We cannot know, but there were strong indications, with the acquisition of Hunslet Foundry in 1810, that his sons were set upon a different path, a refocus on millwrighting. The Jubbs maintained large stocks of timber for their millwrighting business, and the added foundry capacity at Hunslet catered for the trend to cast iron in many industrial applications.[29]

It is difficult to see why millwrights in general would have remained involved at all in textile machine-making after 1800. They would place themselves in competition – for customers and skilled workers – with an increasingly competent and technologically confident new industry, based in the smithy tradition. As soon as iron frames came to predominate, from the 1790s, the millwrights who had already engaged tended to withdraw. Before a wide range of machine tools was available, precision machine components were produced by specialist subcontractors, or subcontractors of subcontractors. Richard Hattersley's Keighley hub, spreading out into a web of neighbouring suppliers with manual expertise in making spindles, rollers, flyers, screws or nuts for spinning frames, is well documented.[30] This was not a comfortable route for millwrights to pursue.

In the generation following Hattersley there are instances of men trained as millwrights moving into textile machine-making. But this was a different world. The industry was then, in the 1820s – the decade that saw the deaths of Hattersley and Matthew Murray, and the venture into flax-spinning machinery by the young Peter Fairbairn – a more strategic affair. Within a maturing industry, a young man could look around and spot opportunities. A new organizational possibility had arisen: the regular arrangement of machine tools on factory floors, alongside integral foundries and forges, and the spectacle of machines making machines. Like his older brother William, Peter Fairbairn had been apprenticed as a millwright, but afterwards he concentrated on the potential in textile-engineering's new model.

From a textile engineer's perspective, millwrights produced the context in which machinery was used. The millwright model transferred into the new factory-based operation. Millwrights designed, built, and installed heavy traditional machinery, such as fulling stocks, and also prime movers and power systems, including steam engines. They acted as main contractors in factory-building, which sometimes (early in the cotton factory boom, *c.* 1800) involved

---

28 WYAS Leeds, 6257/1-2, ledgers 1787–94; *Leeds Intelligencer,* 14 Apr. 1789; 8 Nov. 1791.
29 *London Gazette,* 11 March 1809, 16236, p. 333; *Leeds Mercury,* 29 Sept. 1810; Cookson, 'Hunslet Foundry', 162–3.
30 WYAS Bradford, 32D83/5/3, day book from 1811; 32D83/2/2 (sales ledger 1809–18). See Cookson, *Age of Machinery,* App. 1.

commissioning joiners and smiths to construct machinery *in situ*. The long-established demarcation between millwright and machine-maker, the line where transmission met production machinery, had not moved.

A view gaining some currency at this time was that the status of millwright had been usurped by many, relatively untrained, claimants to the name. They took semi-skilled factory work minding machine tools, it was said, and swamped the market, to the detriment of old-style journeyman millwrights.[31] There may be some truth in this as a factory phenomenon, but it was certainly not the case for highly skilled northern millwrights, whose ever-increasing technical abilities were in great demand. Peter Fairbairn's employees on the machine production line at his Leeds factory (numbering 550 in 1841) were largely semi-skilled machine operatives engaged in repetitive tasks, plus a 'corps of blacksmiths'.[32] These were not millwrights.

But millwright work was far from homogeneous, in either content or structure. Certain discrepancies in how work was described can be explained by substantial regional differences within the trade. Millwrighting in London appears to have had unique features, and was not the force in the mid-eighteenth century that it afterwards became.[33] Campbell's *The London Tradesman* classed millwrights with millers, saying only that

> These two Tradesmen are better understood in the Country than in the City, tho' there are some who live in the City and are concerned in Mills in the Country. The Mill-Wright is an ingenious and laborious Business, in which there is a great Variety, according to the different Principles upon which the Mill is constructed, but the Wages given to Journeymen is no more than that of a common Carpenter.[34]

Waller, Campbell's exact contemporary, took a broader geographical view and is more expansive about the millwright of 1747:

> Their trade is a branch of carpentry (with some assistance from the smith) but rather heavier work, yet very ingenious, to understand and perform which well, a person ought to have a good turn of mind for Mechanics, at least to have some knowledge in arithmetic, in which a lad ought to be instructed before he goes to learn this Art; for there is a great deal of variety in mills, as well as in the Structure and Workmanship of them; some being worked by

---

31 Thompson, *Making of the English Working Class*, 271–2; Moher, 'London millwrights', esp. ch. 1.
32 Cookson, *Age of Machinery*, 237.
33 Moher, 'London millwrights', 7.
34 Campbell, *The London Tradesman* (1757), 322–3, in the appendix to his third edition. This should not be confused with an entirely different book, also first published in 1747, by Waller. The British Library catalogues them separately and correctly.

horses, some by wind; others by water shooting over, and some by its running under; And why not in time by Fire too, as well as engines?[35]

Here are important points. The context makes clear that, when Waller talks of 'Mechanics', he means a branch of mathematical knowledge, and not a worker in metal. And in contrast with Campbell's somewhat dismissive remark about the wages of journeymen millwrights, with the implication of a low-status trade, Waller continues:

> They take with an apprentice 5 or 10 l. [£] work from six to six; and pay a journeyman 12 or 15 s a week; but 50 or 100 l. worth of timber, and 50 l. to spare will make a Master of him.

In other words, the premium paid for a millwright's training was £5 or £10, after which he might work as a journeyman – that is, an employee paid by the day. Yet the other option, reasonably attainable by someone of this class, was to set up in business for himself with £50 to £100 on materials, and a £50 float to cover initial revenue costs.

Moher is also clear that eighteenth-century London millwrights were not metalworkers. He discovered, too, a significant contemporary description by John Smeaton, who employed London's finest millwrights. Smeaton thought the average millwrights there

> very good ones in regard to the mechanical parts, that is the making of wheels, cogs and rounds and in all the small machinery for cleaning and facilitating the operations of cleaning and dressing flour, but as mill architects or engineers, that is as to the great outlines and proportions of the work, they are very strangely ignorant.[36]

Note Smeaton mentioning a set of wooden pieces as 'mechanical' parts. His meaning is moving parts, and he describes a machine made primarily of wood. These were not metal components.

Moher's late-eighteenth-century London millwrights emerge as widely skilled and knowledgeable, yet wedded to 'traditional "rule of thumb" methods' and techniques which appeared inadequate to meet the new demands for better energy sources. Their deficiencies became an issue, 'particularly as they were determined to assert their traditional monopoly'.[37]

William Fairbairn's widely quoted portrayal of 'the millwright of former days' has been influential. It is, though, problematic. Many who quote it

---

35 Waller, *General Description of All Trades*, 151.
36 Moher, 'London millwrights', 31, quoting the Institution of Civil Engineers Archives, Machine Letters of John Smeaton, vol. 2, letter 18 Feb. 1786.
37 Moher, 'London millwrights', 31.

have read a version heavily edited by Pole, his biographer, published in 1877 after Fairbairn's death. The original text of Fairbairn's *Treatise on Mills and Millwork* (1861) was considerably longer. Fairbairn presented his millwright employees as obstinate and confrontational, and objected to their mutual societies and insistence on their rights. The oft-cited accolade to an earlier breed of millwright was in fact the prelude to a comprehensive swipe at his own journeymen. That is hardly surprising. But perhaps more interesting is that he went on to find extensive fault with the older type, those upon whom he had just lavished such praise. They could be dissipated, pig-headed, neglectful, undereducated, reckless, and vain.[38]

Fairbairn's contradictions – commending the eighteenth-century millwright while on the same page unable to resist criticizing his shortcomings – should sound a warning. The famous portrait of a millwright is certainly overwrought, and that may be because Fairbairn had little or no direct experience of such a person. Fairbairn had indeed started out as a millwright, but a colliery millwright trained on Tyneside in the early years of the nineteenth century. Arriving in Manchester only in c. 1816, when in his mid-twenties, Fairbairn had not witnessed the – very different – working arrangements during engineering's foundational years in the textile districts, two decades or so earlier. He makes no comment on that. Nor does he make any reference to millwrights as machine-makers.[39]

But others have done so. Musson and Robinson unfortunately perpetuated the myth about what millwrights did, to the extent of subsuming them into one trade with machine-makers. Their index does not separate out millwrights, but rolls them into 'Engineering, mechanical (machine-making, machine-tools, mill-work etc.)'.[40] Thus an important specialism is dismissed, and their own evidence about well-known Manchester millwrights Thomas C. Hewes and Peter Ewart, which could underline the importance of a grounding in theoretical mechanics, is given no weight.[41]

In summary, then, the millwright's trade was never obsolete, but it adapted, as it always had, to new times and technologies. For very many millwrights this meant switching to (what we now term) civil engineering and factory-building. They must readjust to changing power sources and transmission systems, and deal with structures made of cast iron rather than primarily of wood.

---

38 Fairbairn, *Treatise on Mills and Millwork*, preface to 1861 edition, given in 3rd ed. on pp. ix–xiv; Pole (ed.), *Life of Sir William Fairbairn*, 26–7. Smiles, *Industrial Biography*, 309–14, describes Fairbairn's difficulties with unions when he first sought work in London before settling in Manchester.
39 In BDCE, vol. II.
40 Musson and Robinson, *Science and Technology*, 513.
41 Musson and Robinson, *Science and Technology*, 98–9.

This is quite straightforward. Why, then, has it been so difficult to settle? The answer lies in what is understood by 'mechanic' or 'mechanical'. The lore of the trade, passed down for as long as there had been millwrights, had become insufficient. Particularly at higher, professional, levels, the role of millwright had always possessed an intellectual dimension, underpinning the physical aspects of carpentry, earthworks, and the rest. Mathematics was becoming essential, right across the trade.

In this tradition, Smeaton, to take a particularly impressive example, published his waterwheel guidelines in 1759. These covered the *principles* of mechanics and, like other eighteenth-century references to mechanics, were in fact referring to the *branch of mathematics*. They did not mean engineering in metal.

And, as shown above, to an eighteenth-century millwright, 'mechanical' parts were almost always made of wood. 'Mechanism', wrote William Fairbairn in 1861, 'may be defined as the combination of parts or pieces of a machine, whereby motion is transmitted from the one to the other.'[42] He did not say that this was necessarily a thing constructed of iron, or even of metal.

And while after 1820 the likes of Fairbairn specialized in the theory and design of structural iron – *cast* iron – millwrights were not workers in malleable – *wrought* – iron. Whatever that word 'mechanic' suggests, wrought iron remained the province of the smith, and the smith's industrial descendants.

---

42   Fairbairn, *Treatise on Mills and Millwork*, 12.

5

The reach of science

May God us keep
From single vision and Newton's sleep.

William Blake, letter to Thomas Butt, 22 Nov. 1802

In stories of industrialization, the individuals capturing attention are generally ones involved in producing something original. They are noteworthy because they were inspired somehow to move beyond the settled practice of the time.

Settled practice, though, is the context which had informed them, and in which they lived and worked. Just as social status very much determined someone's occupation and education, so too it decided points of access to the institutions interested in natural philosophy or technologies. This has to be kept in mind when discussing the reach of knowledge – scientific, industry-based, or otherwise.[1]

It was customary in skilled manual trades for son to follow father, subject to ability and an available place in the business. The professions, too, generally ran on succession, though John Smeaton showed that not every son wanted to be a lawyer, just as Robert Louis Stevenson proved that building lighthouses was not for all. So there were general patterns, and notable exceptions. Family dynamics clearly had a part in determining how businesses approached innovation. So too did attitudes to risk, and availability of resources.

Such context is important when considering how science was viewed by industry, especially in rapidly evolving sectors. Partners and business owners could be very open to new ideas, though others appear to have been unwilling or unable to initiate change. Pacey has suggested 'scientifically-minded industrialists' as a type, which suggests an obverse, industrialists not so-minded.[2] But channels ran in both directions. The examples of William Lewis and some important figures among the 'galaxy of Scotsmen' show scientists with strong interests in industrial processes, which they used as a setting for research and experimentation.[3]

---

1 See ch. 4, 'Class and its consequence'.
2 Pacey, 'Emerging from the Museum', 458.
3 For Lewis, see ch. 3; for Scottish influences, ch. 7, 'A galaxy of Scotsmen'.

None of this was neat. As technological change accelerated at speed in the decades before and after 1800, processes were in flux, and remarkably so in the manufacture of iron, engines, and machinery. With adaptation and adoption working on many fronts, change was disorderly. The relationships between natural philosophers and industrialists, it is claimed, were fundamental to working through these technological problems. That is quite likely, though the meeting of science and industry may not have occurred quite in the ways historians have tended to assume.

The focus has to be upon what learned societies, Scottish universities, and public science events delivered for industry. Was usable scientific information diffused to places where it was needed? And converted to practical effect in industrial settings? Mid-eighteenth-century London-based learned societies could, as already established here, be some steps removed from practicality.[4] But the closer business and personal relationships of a provincial industrial town – surely these offered a more promising seedbed in which to float new ideas? Had there been progression from Lewis, and his interest in the Society of Arts, into regional societies – and if so, what were those local groups, and who had access to them?

Joseph Priestley in Leeds and Birmingham

Joseph Priestley's experiences reflect some of these complexities, his career an illustration of varied regional scenes and strands of connection in science and industry.

Before 1800, a handful of provincial 'scientific' formations existed. The earliest were ad hoc, only later taking institutional shape as local philosophical societies. The Leeds group was at most a loose circle, not formally constituted.[5] Priestley, William Wood, surgeon Hey, and John Smeaton, though – here were four eminent men of science, three of them FRS, in association on some level. But all were heavily committed professionally, Smeaton was frequently absent, and Wood arrived in Leeds to replace Priestley as minister at Mill Hill Chapel. So the idea that this circle had industrial influence as a group seems far-fetched, and there is no clear evidence that it did.

Priestley's most famous discovery was made in Leeds. By 1770, he had discontinued studies on optics, to pick up 'some of Dr [Stephen] Hales's inquiries concerning air'. He began exploring new applications for carbon dioxide, then called 'fixed air', which was a by-product of the brewery next door to his first Leeds lodging. Priestley's paper in the Royal Society's *Transactions* two years

---

4 See ch. 3, above.
5 See above, p. 88.

later, 'Observations on different kinds of air', won him the society's Copley medal. It was of exceptional significance: he isolated and identified new gases and processes, and shared details of his apparatus and experimental techniques so that others could build from his findings. The conclusions, though, were problematic, and not resolved during his stay in Leeds. He had apparently learned more from the brewery than it learned from him.[6]

Given Priestley's eminence, writers have been inclined to embroider his wider connections and influence. Joseph Dawson (1740–1813), Unitarian minister, ironmaster, chemist, and mineralogist, has been cast as Priestley's protégé. Dawson, who had been baptised at Mill Hill, left Leeds for Daventry Academy before Priestley's arrival. The two shared interests, circled around the same institutions, and may have met. But Dawson's path diverged from ministry into industrial pursuits and, like his friend David Mushet in Glasgow, he had the frustration of advocating a scientific approach to other ironmasters and having it fall on stony ground.[7]

Priestley had left Leeds for Wiltshire, and the support of an aristocratic patron. In 1780 he returned to urban life, in Birmingham, where he famously joined the Lunar Society, then at its most active. This, the best known of late-eighteenth-century scientific societies, met from c. 1765. It was a select and informal group, never more than fourteen in number and over the years boasting ten FRS. A prime mover (appropriately) was Matthew Boulton, along with Dr Erasmus Darwin, physician and Linnean botanist; Josiah Wedgwood was an early member; James Watt, in contact with Boulton from the mid-1760s, arrived in Birmingham in 1774. This was an exclusive gathering, the 'Lunaticks' including medics, industrialists, dissenting ministers, and others, all of the professional class or gentry. The society's shared interests included aesthetics alongside scientific, medical, and other matters.[8]

In Birmingham, Priestley took a turn into applied science. While the dispute with Lavoisier about the doctrine of phlogiston endured, Priestley's earlier scientific originality did not. Rather he channelled his expertise in gases to the service of industry. In rejoining the ministry, he accepted financial support from his wife's brother, the ironmaster John Wilkinson, to whom he was close. Priestley advised both Watt and Wilkinson on steam interaction with iron, Boulton on the cost and elasticity of newly discovered gases, Wedgwood about airs entrapped in ceramic clays, and William Withering on generating hydrogen

---

6 Schofield, 'Joseph Priestley', DNB; Wykes, 'William Wood, 1745–1808', DNB.
7 Pacey, 'Emerging from the Museum', esp. 456–8, 461, 464. Dawson's close associate Joseph Priestley was a different person, a canal surveyor of the same name. For Dawson, see ch. 6; Mushet, ch. 7.
8 Marsden, *Watt's Perfect Engine*, ch. V; Uglow, 'Lunar Society of Birmingham (act. c.1765–c. 1800)', DNB (rev. 2017: https://doi.org/10.1093/ref:odnb/59220); Ó Gráda, 'Did Science Cause the Industrial Revolution?', 226.

cheaply for balloon flight. Withering, a physician and botanist, was also one of the Lunar Society, a Midlander and alumnus of the University of Edinburgh, and had been taught by (among other notables) Joseph Black.[9]

That observation about Withering shows something about the nature of personal and institutional connections, and the transmission of information. Scientific discoveries were only one part, for it was also knowledge and ideas that were shared in those monthly dining clubs and public lectures. In associations between the philosophical societies themselves, and connections between their individual members, institutional and business networks were very evidently at work.

Links of education and religion are also obvious, with continental and Scottish Enlightenments, and English dissenting academies, influential. Some in this web of connection, most famously Priestley and Thomas Paine, had come to see the revolutionary politics of France and North America as a most urgent concern. Priestley left Birmingham in 1791 after a 'church and king' mob wrecked the house Wilkinson had built for him, destroying Priestley's library, laboratory, and papers, along with several other houses and meeting houses. On this event, the Derby Philosophical Society sent a letter of sympathy, which also exhorted Priestley to 'leave the unfruitful fields of polemical theology' to concentrate on science. With thanks, the minister responded: 'Excuse me, however, if I still join theological to philosophical studies, and if I consider the former as greatly superior in importance to mankind to the latter.'[10] He was consistent in viewing politics and science as part of his theological vocation. Arriving at the truth required disputation, for it would appear out of 'the conflict of contending ideas'.[11] There was no consensus on this, even among Unitarians and their associates. Debate continued.

Philosophical societies

The letter to Priestley itself generated controversy within the Derby society. It had been approved unanimously at a regular monthly meeting, but certain members not present took exception. To one objector, the secretary responded that as the letter did not refer to Priestley's political views, and indeed encouraged him

> to decline those theological controversies which seem to have provoked the vengeance of his adversaries, it was conceived that no man of a liberal mind would object to congratulating him on his escape from the violence of an

---

9 Schofield, 'Joseph Priestley', DNB; Harris, 'John Wilkinson', DNB; J.K. Aronson, 'William Withering (1741–99)', DNB.
10 Musson and Robinson, *Science and Technology*, 197.
11 Schofield, 'Joseph Priestley', DNB.

enraged mob; and that there could be no member of a *Philosophical* Society who did not regret the demolition of his valuable laboratory and manuscripts.[12]

Clearly, individuals joined this and similar societies for a variety of reasons, and with different expectations. While 'a liberal mind' was of course desirable, even assumed, there were political and religious differences, and members' interests, motivations, and aspirations ranged widely. The Derby malcontents, it appears, were among 24 members, of a total of 37, who lived outside the town. Some were probably inactive, or interested only in social opportunities. Such ventures were patronized by intelligent gentry and clergymen, though maybe for form's sake. Regular meetings attracted the most committed members, those interested in natural philosophy and radical politics, and even scientifically minded manufacturers. Clearly there was a range of types. Gatherings cannot have passed without argument, verging on strong disagreement. In general the manners of their class enabled them to rub along politely, though at times their differences burst out into public view.

The Lunar Society stood as a gold standard, unlikely to be elsewhere replicated. It was by invitation only, tiny, and without a formal membership structure. The Derby Philosophical Society more closely resembled a model emerging in some other industrial towns. It was established by Erasmus Darwin when his move to Derby in 1782 removed him from close association with the Lunar Society, of which he had been a founder member. At the start of 1784 his 'infant philosophical society' had attracted seven members, and Darwin proposed to Matthew Boulton that there be occasional joint meetings, though 'we do not presume to compare it to your well-grown gigantic philosophers at Birmingham'.[13] This was more than superficial politeness: in deferring to the Lunar men, Darwin acknowledged that his new formation in Derby was of a different brand.

Darwin was central to the circle of Midlands philosophers. He had a strong link too to the Manchester Literary and Philosophical Society, of which he was first made an honorary member, and then a fellow from 1784. The new associations emulated each other, while also drawing on the example of literary societies and subscription libraries. Each reflected the character of its own locality. Darwin referred to Freemasonry when suggesting that the learned societies build reciprocal connections and welcome guests from other towns. But Freemasonry was highly centralized, while the Lit. & Phils (as the societies became colloquially known) had no formal network, and interconnections were more personal.

---

12   Musson and Robinson, *Science and Technology*, 197–8. Italicized as the original.
13   Musson and Robinson, *Science and Technology*, ch. 4 on the Derby society, is the main source of quotations here.

Under Darwin's energetic leadership, the Derby group thrived, losing impetus after he died in 1802. But there were other notable members. One was Darwin's friend Robert Bage (1728?–1801), his former partner in a slitting mill at Wychnor (Staffordshire). Bage was also a novelist, a political radical hounded by the authorities, and involved in a number of business enterprises.[14] Another member was the son of an early Arkwright partner, William Strutt (1756–1830), a radical who knew Thomas Paine and was heavily influenced by Darwin. Strutt was involved in planning the first fireproof factory, Ditherington Mill, Shrewsbury, in 1796–1800, alongside Bage's son. Charles Bage, the main designer, joined John Marshall of Leeds and the Benyons as partner in this flax-spinning venture. The columns were to be of iron, and Bage decided also to substitute iron beams for timber. Paine had some influence on the choice of structural ironwork for the Wearmouth bridge, Sunderland, which opened in 1796.[15] Whatever the extent of this, there was demonstrably a breadth of personal connection at work. While some of these families were already acquainted, there is no doubt that the brotherhood within learned societies brought talented people together to speculate, learn, and scheme. The Derby society, it is said, had particular interests in applying scientific research to commercial needs.

Darwin's correspondence is enlightening about his own approach to 'public' science. Here was a man capable of holding his own in any forum, whether scientific, professional, or social – his second wife was daughter of an earl. Nor did he fear espousing unpopular opinions, and he was a noted conversationalist.[16] What, then, is to be made of this: in 1789 Darwin wrote to James Watt, asking for information about Watt's steam engine, but presented in a particular form: 'such facts, or things, as may be rather *agreeable*; I mean gentlemanlike facts, not abstruse calculations only fit for philosophers'.[17]

What constitutes a 'gentlemanlike fact' has been the cause of much subsequent head-scratching. Musson and Robinson hint that it means something unsuitable for 'the ladies'. But the progressive Darwin would hardly have tried to exclude women from learning. Musson and Robinson themselves cite an example of Darwin bringing young women into scientific conversations.[18] Darwin wrote

14 McNeil, 'Erasmus Darwin (1731–1802)', DNB; Kelly, 'Robert Bage (1728?–1801)', DNB; https://www.revolutionaryplayers.org.uk/a-biography-of-robert-bage; Goss, 'A radical novelist in eighteenth-century England'. Bage's novels fictionalize the scientific interests, ideas, and values of the members of his and similar societies, across Britain and Europe.
15 'William Strutt FRS (1756–1830)', BDCE; 'Charles Woolley Bage (c. 1752–1822)', BDCE; Giles and Williams (ed.), *Ditherington Mill*, 1–2; Rimmer, 'Castle Foregate Flax Mill'; Newman and Pevsner, *Shropshire*, 587–90; Cardwell, *Fontana History of Technology*, 170; Cookson (ed.), *History of Co. Durham*, V, 241–4.
16 McNeil, 'Erasmus Darwin (1731–1802)', DNB.
17 Birmingham Library, MS 3219/4/80B, Darwin to Watt, 20 Nov. 1789.
18 Musson and Robinson, *Science and Technology*, 192.

A Plan for the Conduct of Female Education, in Boarding Schools (1797) as advice for his two daughters born outside marriage, who planned to open a school. While he did not advocate an identical curriculum across the genders, his ideas on the education of young ladies, including in science, were ahead of their time.[19]

'Gentlemanlike' was in fact a lighthearted message to Watt, an old friend, that a simplified explanation of the steam engine was required here. It must be accessible to people outside Watt's usual circle, those without the Lunar men's scientific understanding, or the technological specialisms of his own industrial contacts. Whether Darwin intended his presentation for publication, or for the Derby society, is uncertain. Whichever, it was part of Darwin's mission to connect with men, and perhaps even women, who wanted to learn more about the basic principles of Watt's engine. Their interest may be primarily social, but it could also be serious, and was to be encouraged. Darwin fully understood how his Derby group differed from the erudite Lunar men.

There is a varying richness and texture about the societies. Their ethos rested on the quality of leadership, and levels of learning and influence among the more talented members. By their nature, groups brought together the better educated, so good records were maintained of their membership and activities. Local distinctiveness is evident. Establishing any direct industrial influence is less straightforward. And if this did occur, can it be credited to the societies themselves? They assembled some great minds, though in many cases those great minds were already acquainted, and it was they who established the philosophical meetings.

Of all provincial societies, the most interesting examples were Manchester and the Midlands set – although 'set' suggests something more collective than was actually so. Any links between the groups were informal, partly because they were differently constituted. The Lunar – élite, exclusive, somewhat irregular – was not in the 'Lit. & Phil.' mode. Derby and Manchester conformed to a wider pattern, comprising intellectual leaders of a provincial town. Even so, there were substantial differences between the two. In neither case can it be said that industrial advance as such was on their agenda.

Bruton's investigation of the Shropshire Enlightenment is instructive here. He shows that technological innovation originating or adopted within Shropshire owed much to intellectual exchanges between serious thinkers. Yet this happened externally, not within a society forum. Before the 1830s this most innovative of eighteenth-century counties did not have a philosophical society. The local channel must have been personal communication, written or spoken. That applied to Charles Bage and William Strutt, in developing a pioneering approach to structural ironwork at Ditherington; and, as a general *modus*

---

19 McNeil, 'Erasmus Darwin (1731–1802)', DNB.

operandi in the course of their work, to John Wilkinson and the ironmaster William Reynolds. Reynolds' wide interests included minerals, electricity, and chemistry. He had spent time with Joseph Black, and was dedicated to applying science to his industrial pursuits. He corresponded with James Watt, who wrote about his pleasure at having 'at last got a fellow labourer in the pneumaticle vineyard. I mean in the chemical part'. Watt sent three pages of notes on chemistry, inviting Reynolds to visit for more discussion. Richard Crawshay thought that Reynolds had 'more metalurgie and chemical skil than any other of my friends'.[20]

Looking more widely at the Lit. & Phil. phenomenon, Bruton found that a large proportion of members were medically trained. Adding in the social element, the clerics, gentry, and women members not conversant with manufacturing, this does not suggest the Lit. & Phil. was a place for much discussion about industrial technology. Remember too that Shropshire was not unique in being without a philosophical society in those years: before 1800, neither did major industrial towns like Newcastle, Sheffield, Bradford, and (in any meaningful sense) Leeds.[21]

The Manchester Literary and Philosophical Society
The Manchester society was a different matter, with ambition, strategy, and excellent connections rooted in nonconformist colleges. It would become a leader in science, in England considered second only to the Royal Society.[22] The Manchester Lit. & Phil. started in a formal sense in 1781, building from a more casual group which had met for several years. Its first historian traced the society's beginnings to the dissenting academy which opened in Warrington in 1757. This, and similar institutions such as Daventry and Kendal, accepted bright sons of nonconformists whose religion excluded them from English universities. The wealthier could choose Scotland or the continent for higher education, but the academies were worthy alternatives. As a tutor among the small group of intellectuals delivering a liberal education in Warrington from 1761 until his move to Leeds, Priestley had been enabled to pursue his growing interests in science.[23]

---

20 Bruton, 'Shropshire Enlightenment', 102–6, 152–3, 221. Like his grandfather, Abraham Darby II, Reynolds did not generally patent his innovations: Bruton, 'Shropshire Enlightenment', 123.
21 Bruton, 'Shropshire Enlightenment', 218–20, 223–4, 153. The Leeds society founded in 1819 was the Phil. & Lit. For Bradford, see Morrell, 'Wissenschaft in Worstedopolis'.
22 Greenaway, 'Thomas Henry', DNB.
23 Smith, *Centenary of Science in Manchester, passim*; Cardwell, *Organisation of Science*, 20–4. The academies were particularly associated with Unitarians, though serving a wider constituency. Quakers had their own network of schools.

Dr Thomas Percival MD FRS was instrumental in creating the Manchester society. From a line of physicians in Warrington, Percival was the academy's first student, continuing his education at the University of Edinburgh (where he knew Hume and Robertson) and then in London, Leiden, and Paris. Percival set up practice in Manchester, alongside developing interests in sanitary and factory reform, epidemiology, and medical ethics. He supported the Warrington Academy's move to Manchester in 1785.[24] The philosophical society that he helped build, as president from 1782 until his death in 1804, became a shining example of its type. For this success, Cardwell credits John Dalton. The Manchester society inspired 'a distinguished and very creative school of scientists' to collect around Dalton, and in consequence the town emerged as a centre of European science.[25]

Dalton had attended a Quaker elementary school in Cumberland, though from the age of ten his education was piecemeal. He arrived in 1793, as a largely self-educated professor of mathematics and natural philosophy, at Manchester Academy, later called New College. Dalton was pivotal to Manchester's scientific advancement, and his blossoming as a scientist was enabled by the Lit. & Phil. and its Unitarian founders. Percival was close to the society's first secretary Thomas Henry (1734–1816), and Henry's son William (1774–1836). All three were FRS with medical backgrounds. The Henrys were deeply interested in chemistry, and for a time William Henry was secretary and assistant to Percival, and a significant collaborator with Dalton. The Lit. & Phil. gave Dalton an institutional base, allowing him to leave the academy and focus on science. Thereafter he worked freelance, living frugally. He served as the society's secretary, 1800–9, was then vice-president, and took over as president when Thomas Henry died, continuing until his own death.[26]

When launched in 1781, the ManchesterCed. & Phil. limited its ordinary membership to fifty. In addition, it invited a list of honorary members, among them Lunar men – Erasmus Darwin, Joseph Priestley, Josiah Wedgwood – and an extraordinary range of international talent, including Benjamin Franklin, Monsieur Lavoisier, and Professor Volta of Como. Like Darwin and his Derby society, formalized at virtually the same time, the ambition was to link into wider intellectual circles. Although FRS remained a mark of distinction, this was a time of dissatisfaction with what Cardwell describes as 'the lethargic

---

24 Smith, *Centenary of Science in Manchester*, esp. ch. 3; Nicholson, rev. Pickstone, 'Thomas Percival (1740–1804)', DNB.
25 Cardwell, *Organisation of Science*, 65. Greenaway, 'John Dalton', DNB, also refers to Dalton's relationship with the Henry family.
26 Greenaway, 'Thomas Henry (1734–1816), apothecary and chemist'; 'William Henry (1774–1836), physician and chemist'; 'William Charles Henry (1804–1892), physician and chemist', DNB.

Royal Society and ... the tyrannical rule of its president, Sir Joseph Banks'. This contributed to a flowering of new national and provincial societies.[27]

Cardwell has significant insight into the Manchester group's relations with industry. The approach of Arkwright and similar textile innovators, he suggests, had been almost entirely empirical. The late eighteenth century saw growing interests in science and technological development, particularly evident in northern and Midlands industrial towns. The increase in peripatetic science lecturers is one reflection of this. However – importantly – from an industrial perspective, not too much should be read into that. The lecturers delivered only basic science, and, Cardwell concludes, from their quarter 'there was no indication of the practice of applied science'.[28]

But the potential in science to inform industrial practice had not escaped the Manchester Lit. & Phil. founders. Thomas Henry explained this in his first address to the society, in October 1781: 'The misfortune is that few dyers are chemists and few chemists are dyers.' He continued:

> Practical knowledge should be united to theory, in order to produce the most beneficial discoveries. The chemist is often prevented from availing himself of the result of his experiments by the want of opportunities of repeating them at large; and the workman generally looks down with contempt on any proposals the subject of which is new to him ... the arts of dyeing and printing owe much of their recent progress to the improvements of men who have made chemistry their study. Much, however, remains to be done.[29]

Henry's vision and his frustrations were shared by others, in his group and more widely – contrast that with Dawson and Mushet and their inability to interest fellow ironmasters in chemistry's potential.[30] But Manchester was different. A vision was already defined and understood, and an influential group gathered. The educational structure they built nourished scientific advance. Dalton was first a beneficiary of this, before underlining his place as nucleus of the town's scientific revolution. Perhaps most importantly, the focus was intense upon specific areas of chemistry and in the service of defined local industrial needs. Inevitably, the course of events was not straightforward. The first difficulty was to sustain an educational framework.

The minister Thomas Barnes, of Cross Street Chapel, who with Thomas Percival and Thomas Henry had established the Manchester society, shared

---

27 Cardwell, *Organisation of Science*, 22–3. Banks was president of the RS from 1778 until his death in 1820. The Royal Institution, established in 1799, was more go-ahead and practical, with a focus on science and including women among its patrons: see James, 'When Ben Met Mary', esp. 232–4.
28 Cardwell, *Organisation of Science*, 22.
29 Smith, *Centenary of Science in Manchester*, 78–85, esp. 84.
30 See p. 183

Henry's sentiments, writing in 1785 that 'few of our mechanics understand the principles of their own arts and the discoveries made in other collateral and kindred manufactures'. Barnes, with others, promoted a College of Arts and Sciences in 1783. This offered a broad curriculum, but with chemistry and mechanics taking priority, reflecting their importance to manufacturing. Barnes saw the promise in applied science, 'the happy art of connecting together liberal science and commercial industry'. The college lasted only until 1787, perhaps eclipsed by the new Manchester Academy, where Barnes became the first principal.[31]

Thomas Henry taught chemistry at the college, in evening classes timed to suit workers. In his own career Henry acted out what he was advocating for others, opening a business as manufacturing chemist alongside his apothecary. His magnesia factory was very successful in supporting his family for several generations afterwards. In the field of theoretical chemistry, however, the son outshone his father. William Henry won the Royal Society's Copley medal in 1808 for work on the solubility of gases, becoming FRS the following year. He subsequently turned down an approach to be professor at Edinburgh, citing his commitment to the Manchester business.[32]

The younger Henry also collaborated closely with Dalton, and in fact was considered the better experimentalist. Their friendly arrangement worked to mutual advantage. Henry's law, demonstrating that, at a given temperature, water dissolves the same volume of a compressed gas as of that gas under normal pressure, supported Dalton's development of atomic theory, which drew on the physics of gases and vapours. Dalton published many articles, over thirty of them under the banner of the society's *Memoirs*. Much of his income came from lecturing, and he took industrial commissions, including water quality analysis for a bleaching company. William Henry's goal was to identify practical applications for these discoveries. In this he was very successful. His breakthroughs in understanding gases, especially coal gas and other illuminating gases, created great commercial potential and also promised social benefits.[33]

During this period, other threads were drawn together on the platform of the Manchester Lit. & Phil. The principal industrial focus of Manchester's early scientific community was turned towards two sectors: gases and textile dyeing and bleaching. Both subjects had preoccupied enlightenment philosophers, and the industrial possibilities were widely recognized. Encouraged by Thomas Henry, the College of Arts and Sciences focused on the scientific

---

31 Cardwell, *Organisation of Science*, 22–3; Wykes, 'Thomas Barnes (1747–1810), Presbyterian minister and reformer', DNB.
32 Greenaway, 'William Henry', DNB; Clow and Clow, *Chemical Revolution*, 606–7.
33 Greenaway, 'John Dalton'; 'William Henry', DNB; Clow and Clow, *Chemical Revolution*, 189; Cardwell, *Organisation of Science*, 24.

development of chlorine bleaching, with immediate practical purposes in view.[34] To pursue chemical knowledge for industrial uses with such single-mindedness was exceptional, perhaps unique in the context of a philosophical society. But Manchester's was an extraordinary Lit. & Phil.

And that was not all, for James Watt's scientific interests brought him and his son into this Manchester circle. The common ground was chlorine bleaching, and afterwards a mission to develop gas plant.[35] The younger James Watt spent about four years in Manchester, participating very actively in the Lit. & Phil. and investigating both of those topics. He worked for Taylor & Maxwell, fustian manufacturers, from 1788. In that same year, Dr Charles Taylor and others ran an experiment there to bleach, print and calender a piece of cotton in three days. The result was so impressive that Dr Thomas Cooper, a friend of the younger Watt, immediately established a bleach works based on this technique, near Bolton.[36]

In encouraging his son towards Manchester, perhaps Watt had belatedly come to see that the future of steam engines lay in Lancashire, not Cornwall.[37] But the son, already highly educated and as iron-willed as his father, was not conspicuously selling engines. Instead he continued a rigorous course of self-improvement, growing close to Percival and the Henrys. Appointed co-secretary of the Lit. & Phil., Watt jun. published two papers on minerals in its *Memoirs* for 1790. He followed lectures by Thomas Henry and Charles White at the College of Arts and Sciences, presented his own translations of continental scientific works at society meetings, and assisted James Keir with an article on bleaching for Keir's *Dictionary of Chemistry* (1789). Charles Taylor, later secretary of the Society for the Encouragement of Arts, Manufactures and Commerce, wrote in the same volume, on dyeing and calico printing.[38]

It is much more likely that the young Watt – then nineteen – was dispatched to Manchester as a part of his education, to learn more about those important investigations in progress. There are signs that Boulton & Watt looked to diversify ahead of 1800, when the patent central to their business model would expire. Watt the elder was absorbed by chlorine bleaching as early as 1786, the date he visited Berthollet in Paris. He was among pioneers introducing to Britain the concept of chlorine as a bleaching agent. And then, in 1788, came Watt's own bleaching trials in Glasgow, and we know this because he immediately communicated the news to Thomas Henry. If Watt had any inkling of patenting chlorine bleaching, it was futile, because the practice (soon improved

---

34 Clow and Clow, *Chemical Revolution*, ch. IX, esp. 189.
35 See above, p. 90.
36 Robinson, 'James Watt, engineer and manufacturer (1769–1848)', DNB; Clow and Clow, *Chemical Revolution*, 189. Bolton had more convenient access to coal.
37 The view of Robinson, 'James Watt', DNB.
38 Robinson, 'James Watt', DNB; Clow and Clow, *Chemical Revolution*, 189.

into a more manageable liquid version) was already spreading in Scotland, in Aberdeen and Dunfermline, as well as in Lancashire.[39]

Manchester Lit. & Phil. members made other advances in the theory and practice of chlorine bleaching, among these a significant paper by Theophilus Lewis Rupp on developments in cloth bleaching. This appeared in its *Memoirs* in 1799. Rupp, partner in a Manchester cotton-spinning business, was absorbed in chemistry and is also credited with the mechanical concept of hypocycloidal motion.[40] A continental breakthrough in dyeing, a rediscovery of the elusive 'Turkey Red', had generated similar enthusiasm. Full practical detail of the dye was shared by Thomas Henry in a lengthy presentation to the society in 1786, published in the following volume of *Memoirs*. Charles Taylor set up a new works in Manchester, one of several manufacturers to take advantage of this newly available colour.[41]

In the Manchester context, then, important links with industry were building, and that included philosophical experiments of value to both manufacturing industry and scientific theory. So close-knit were the individuals involved, so wide-ranging their interests, so intense the rate of activity, that it is not always possible to discern in which direction the benefit was moving. This is one important finding. It was also very specific – to the individuals, the local industrial context, and this particular branch of science, chemistry. There was nothing quite like it in any other philosophical society, and the uniqueness of the Manchester phenomenon is to be underlined.

Alongside the dyeing and bleaching interests, a second preoccupation grew within the Lit. & Phil. The same Manchester group, with the two James Watts, followed Priestley, Lavoisier, and other European scientists into the study of gases. For Dalton and his collaborators, it was about linking newly discovered gases into atomic theory. But also in Manchester, again centring upon the Lit. & Phil., came a mounting curiosity about practical applications. The idea of coal-gas lighting was not new, but an innovation must find its time and market. Manchester was then a town ahead of its time, and demand for a novel lighting system was generated by a new style of factory, multi-storey and fireproof. The possibilities for extending the system into street and domestic lighting quickly became apparent.[42]

---

39 Clow and Clow, *Chemical Revolution*, 186–9; Mantoux, *Industrial Revolution*, 250–1.
40 Rupp, 'On the Process of Bleaching'. Rupp had published a response to Priestley's air experiments in *Memoirs of the Manchester Lit. & Phil. Soc.* in 1790. See Anderson, 'Industry and Academe in Scotland'; Cookson, *Age of Machinery*, 207; *London Gazette*, Aug. 1801, 908.
41 Henry, 'Nature of Wool, Silk, and Cotton, as Objects of the Art of Dying'. See also Clow and Clow, *Chemical Revolution*, 216–20; Mantoux, *Industrial Revolution*, 250–1.
42 Clow and Clow, *Chemical Revolution*, 426–32; 449–55.

How to explain Boulton & Watt's involvement? The answer lies with the company's most talented and innovative engineer, William Murdock. Murdock's greatest achievements, the most outstanding of these a package of gaslighting technology in the 1790s, were accomplished in spare-time experiments during two decades of exile in Cornwall. His creativity had been sidelined by his employers – for reasons to be discussed, and which include his social standing. Murdock's story is itself most illuminating, a revelation of how this renowned business actually managed innovation.

Murdock was the most loyal of workers, employed by the Birmingham company from 1777 until forcibly retired in 1830. From training as a millwright with his father in Ayrshire, he became Soho's most skilled engine erector. His prolonged posting to Cornwall was a result, and he was highly recommended by Watt to the company's customers. But as an innovator, Murdock outshone, and consequently irritated, Watt. It is now accepted that the famed sun-and-planet gear, a late change to Boulton & Watt's 1781 steam-engine patent, was inspired by Murdock and replaced Watt's impractical ideas.[43] Certainly many of Murdock's later proposals were resisted by Watt. Once they had their patent, producing increasingly efficient engines was less important to Boulton & Watt than protecting their existing system. The company's model was to install their apparatus so that it would run uninterrupted, and to ruthlessly defend their monopoly against piracy. Murdock's creativity undermined this status quo. An unchanging product maximized profits, and made the real money. Watt's attitude to Murdock was summarized in a letter to Boulton in 1786:

> I wish William could be brought to do as we do, to mind the business in hand, and let such as Symington and Sadler throw away their time and money, hunting shadows.[44]

In Cornwall Murdock turned to his own technological interests, in self-propelled vehicles and machine-tool design. But his principal achievement related to the illuminating properties of gases. Murdock's father had been a renowned innovator, commissioning the Carron works to produce what may have been the first iron-toothed gearing used in millwork, in 1760. The son reached into fields far beyond. Clow and Clow describe 'a genius for chemical experimentation' to match any member of the Lunar Society. Experiments carried out at Murdock's home in Redruth in 1792 subjected different classes of coal, wood, and other fuels to destructive distillation at various temperatures. By 1794, with a larger retort in his yard and about twenty metres of tinned iron and copper pipes delivering coal gas through a window frame, he succeeded in lighting a living room. This attracted some local attention as a

---

43 Birse, 'William Murdock (1754–1839)', BDCE; Griffiths, 'William Murdock', DNB.
44 Griffiths, 'William Murdock', DNB.

novelty, but without serious interest. Murdock proposed to the younger Watt that they take a patent, but Watt was distracted by issues with the steam-engine patents. In fact, the Watts knew of other gaslight initiatives of which Murdock was apparently unaware, trials from the early 1780s. When Murdock returned to Birmingham in 1798–9, news was spreading of similar tests in Paris. Finally Watt became engaged by gas-plant possibilities. From Manchester, William Henry encouraged Murdock to pick up his experiments, which led the engineer to devise new methods of washing and purifying gas, and develop horizontal retorts for gas collection.[45]

The Manchester cotton-spinner George Augustus Lee recognized that gaslighting was not only cheaper and more satisfactory than alternatives, but also enabled round-the-clock factory working, multiplying the owners' return on investment. He steered Murdock towards finding commercial fruition with his discoveries. Lee was on the fringes of the Manchester Lit. & Phil., though not noticeably active there. He had started as a cotton-mill clerk, and rose to managing partner of Philips & Lee, whose cotton-spinning factory opened in Salford in 1791. Lee was also 'imbued with a love of the sciences' and expert in mechanical theory, a skill he deployed when commissioning machinery, 'the finest specimens of perfect mechanism'. Lee was a notably progressive factory owner, technologically at least. His Salford Twist Mill was the second after Ditherington to be cast-iron framed and fireproof, and an early adopter of steam heating. His partners, the Philips family, were active founder members of the Lit. & Phil., and Lee joined at the start of 1790. He was certainly well connected, close to the Watts and Murdock, Strutt and Bage, William Henry, and other illustrious Lit. & Phil. members. Lee first witnessed Murdock's gaslight system towards the end of 1800, enthusiastically adopted it in his own home in 1804, and on the first day of 1806, 50 lights, quickly extended to a 904-light installation, were lit at Philips & Lee's factory.[46]

Boulton & Watt had limited commercial success with gaslighting, and their involvement was short-lived. The company ran trials at Soho, reportedly lighting the factory and counting house by gas for a celebratory event in 1802. The younger son, Gregory Watt, having seen a public display of gaslights in Paris, wrote to urge that Murdock's plan be adopted and the rights secured without delay. Boulton & Watt, with the founders' sons at the helm from 1800, developed a complete package to light a large building. The Philips &

---

45 Birse, 'William Murdock', BDCE; Griffiths, 'William Murdock', DNB; Cossons, *Industrial Archaeology*, 220; Clow and Clow, *Chemical Revolution*, ch. XIX, esp. 426–30. Elsewhere, the Clows, *Chemical Revolution*, 612–15, overstate the Lunar effect.

46 Mason, 'George Augustus Lee (1761–1826), cotton spinner', DNB; Clow and Clow, *Chemical Revolution*, ch. XIX, esp. 426–33, 449–55; Giles and Williams (ed.), *Ditherington Mill*, 57, 64–5; Musson and Robinson, *Science and Technology*, 99–100; Smith, *Centenary of Science in Manchester*, 425.

Lee factory was the prototype, and afterwards a site for large-scale experiments. William Henry, assisted by Lee, also carried out investigations there on preparing and purifying gas. There followed a rush of interest from across the north and in Scotland, including from Benjamin Gott for his Leeds woollen-cloth factory. The savings over candle power were significant, a cost reduction in the order of 70 per cent, plus the financial benefits of 24-hour working. But Gott's gasometer house turned out to be one of a very few. The Soho package, a self-contained model of gaslighting, was almost immediately redundant. Urban demand for house and street lighting quickly reshaped the technology towards using a public supply. Boulton & Watt had misread the market, not foreseeing the advantages of supplying a town from a central source. As this gained ground, the firm took legal action, but the public gas movement was unstoppable. Soho would soon abandon its involvement in gaslighting.[47]

What is to be drawn from all this about the contribution of philosophical societies to industrial innovation? The Manchester society stood alone in its focus on supporting scientific and practical endeavour. Its uniqueness attracted outside interest, so that it benefited from collective efforts to solve specific problems. Specificity was its other great strength. There was a clear and defined focus on chemistry and the application of chemistry. There was also an understanding of the absolute importance of education, and not only for the middling classes.

In contrast, the Lunar Society's members accomplished great things individually, in science or in their own industrial interest. But what is the evidence of group achievement? It was there, yes, but it was abstract. This was a dining club, a talking shop, a place for intellectual growth, to exchange ideas and thrash out principles, to seek intelligent confirmation. In itself, this élite society was emphatically not a crucible of industrial revolution.[48] Murdock's flower blushed unseen in Cornwall for so long that he arrived too late for the Lunar, but had he been around 1780s Soho, he would not have been invited. It was a polite society, a world to which Murdock did not belong.

These societies, the Lunar and the Manchester, with much in common in principle, differed in essence. That is not to say that the Lunars were anything less than serious, resolute, and heavyweight. So too, of course, were those who promoted the Manchester Lit. & Phil. But from its very start, that group was clear in embracing a practical element that was immediate and aspired to be outward-reaching. Thomas Henry suggested in 1781 that certain sciences – he specified mechanics, hydrostatics, hydraulics, and chemistry – would be particularly beneficial for the tradesman, 'to fill up the vacant hours in which [he] can

---

47 Clow and Clow, *Chemical Revolution*, 429–35, and ch. XIX, passim, describing earlier experimentation in coal gas as a source of power; Musson and Robinson, *Science and Technology*, 100, 163.
48 Though see Uglow, 'Lunar Society of Birmingham, (act. c.1765–c.1800)', DNB.

withdraw from his employments' and 'supply him with a kind of information he may turn to good account'. However, such tradesmen did not join.[49] The first lectures offered were on professional topics, lacked relevance to industry, and – perhaps the greatest impediment – were not accessible without a certain degree of education and some ready cash. The social gulf yawned wide.

The Manchester society, though, offered more than good intentions. Its ambition was explicit, the sense of purpose tangible. More than that, an intellectual rigour was injected by the local nonconformist academies, and by the Edinburgh graduates, many of them medical men, strongly represented among the society's leading members. Dalton, an exceptional figure standing at the hub of an exceptional phenomenon, was a constant physical presence in the Lit. & Phil. headquarters, surrounded by a small nucleus of talented experimentalists. This group shared its findings widely and speedily. The society published original scientific papers from the beginning, while its leading lights brought out their own books: William Henry's *The Elements of Experimental Chemistry*, first issued under another name in 1799, was into its sixth edition by 1810; and of course Dalton's *A New System of Chemical Philosophy*, appearing in parts from 1808.

In the small field of provincial learned societies, Manchester was more than a rarity. It had no rival. If mentioned in the same breath as the lofty Royal Society, that probably flattered the RS more than the Lit. & Phil. The initial impetus in Manchester had not necessarily emanated from industry, but that was a direction it picked up, responding to immediate local concerns. Dalton's inspiration lived on, and Manchester Lit. & Phil. would nurture the outstanding talents of James Prescott Joule and Eaton Hodgkinson in the decades that followed.[50]

Manchester Lit. & Phil.'s uniqueness lay in its connectedness. The experimentalists linked closely with scientifically interested manufacturers, fellow members, to address specific industrial questions. The defined agenda worked to everyone's benefit, and results were rapidly disseminated. There were robust links to education. The boundary between science and industrial practice, the balance of gains between those fields – are these even possible to define and quantify? It can be confidently asserted that this was a reciprocal arrangement which rewarded both parties. And there is no question that scientific knowledge gained richly from these close encounters with industry.

---

49 Musson and Robinson, *Science and Technology*, 91–3.
50 Cardwell, *Organisation of Science*, 65.

Public science

Itinerant lecturers are a source of fascination to some of the historians interested in how scientific knowledge was disseminated. But what was the lecturers' impact on their own contemporaries, those with interests in industries which could have benefited from access to science?

This particular hinterland of science calls for some objective reassessment, accompanied by a good measure of scepticism. Lecturers as missionaries of science, carrying learning to the far reaches of provincial ignorance, is of course caricature. The regions, as we have seen, contained plenty of knowledge of their own, particularly about local industrial specialisms. Nor is there an identikit lecturer: like the Lit. & Phils, they did not come standardized.

Public lectures in science began in the late seventeenth century, and ran into the nineteenth. They therefore preceded regional philosophical societies by two generations.[51] Many of the lecturers, including natural philosophers of considerable standing, pursued this work for the stream of income it offered. Even the most famed and respected, men like John Theophilus Desaguliers and John Warltire, could be financially precarious.

What was the general stature of the lecturers? The itinerant teachers were few, and their background and approach disparate. Some of those described as such, it emerges, travelling little or delivering lectures infrequently, were barely 'itinerant'. The pattern, if there was one, also evolved. Certain lecturers on the later-eighteenth-century circuit were themselves provincial, and very conscious of industrial concerns. But overall, presentations and publications were sparse, and thinly spread. Nor could they offer industrial solutions. Even the renowned John Banks, active towards the end of the period and immersed in the dynamic north-west of England, did not present customized 'how to' solutions. His guidance fell into the realm of principles and demonstrations of established technologies.[52] Considering the dynamic context, the speed of technological change at that time – could a public science lecture have addressed specifics about rapidly changing needs across a range of industrial ventures? The question answers itself.

The lecturers' main offer was general concepts. These, particularly opportunities to learn more about maths and physics, could well have assisted an open-minded 'rule-of-thumb' type of millwright. The option was potentially helpful too in building confidence and efficiency in mechanical engineering, then emerging as a discrete trade. There was a growing appetite for mathematics in such quarters.[53] Yet many with the most to gain from such insights were

---

51 Bruton, 'Shropshire Enlightenment', 222–4.
52 Musson and Robinson, *Science and Technology*, 108–9.
53 Cookson, *Age of Machinery*, 174–5.

excluded by cost from attending lectures or reading the books. John Kennedy, then an engineering apprentice at Chowbent, later partner in M'Connel & Kennedy, was enthused by Banks's lectures in c. 1784. He was able to attend only because his employer knew Banks and arranged a local series of presentations. Kennedy shared a five-shilling ticket with another apprentice, taking turns to attend.[54]

John Banks (1740–1805) was active in northern England – Manchester, Leeds, Penrith, and elsewhere – from the 1770s. A solid grounding in mathematics, gained at Caleb Rotheram's dissenting academy in Kendal, underpinned Banks's subsequent career. His strength tests on cast-iron beams for bridges and steam engines were among the first published.[55] Those experiments linked to his consultancy with Aydon and Elwell, who established the Shelf ironworks, near Halifax, in 1794. Banks had strong connections to Manchester, with the Lit. & Phil. and the college, where he taught.[56] He was also famed for three books, all running to several reprints. A final reissue of *On the Power of Machines* came in 1829, almost twenty-five years after Banks's death.[57]

The first volume, to accompany his public lectures, ranged over many scientific topics. The second and third differed, their focus specifically engineering. They contained explanations of general principles of mechanics and hydrostatics. But they were not handbooks and not to be mistaken for engineering solutions. Rather, as a 'man of science', Banks might offer ideas to engineers, or at least share some limits of what was possible.[58]

Banks's serious purpose, and his own view of his role, stand in sharp contrast to some contemporaries who are described as public lecturers. In 1807, Dr Thomas Chalmers, a Scottish clergyman, attended a London lecture and exhibition with scientific pretensions.

> The lecturer, Mr Winsor, is a mere empiric, not a particle of science, and even dull and uninteresting in his popular explanations. The Londoners listened with delight, and I pronounce the metropolis the best mart of impudence and folly.[59]

54 Musson and Robinson, *Science and Technology*, 100–1, 108.
55 Thanks to John Suter for sharing a draft article about John Banks's career. Suter proposes a possible 'Yorkshire school of bridge engineering' deriving from Banks's influence, based on design features of surviving bridges at Scarborough Spa, Newlay (Horsforth), and Walton Hall.
56 Musson and Robinson, *Science and Technology*, 93, 100–4, 107–9; Suter draft article. See also Soares, 'John Banks'.
57 *An Epitome of a Course of Lectures on Natural and Experimental Philosophy* (1775, reprinted 1789, 1794, 1800); *A Treatise on Mills* (1795); *On the Power of Machines* (1803).
58 For descriptions and interpretation of Banks's writings, see Soares, 'John Banks', esp. 24–6.
59 Clow and Clow, *Chemical Revolution*, 434.

Yet Frederick A. Winsor, despite his own shaky scientific credentials, had taken out patents for gas-making apparatus, from 1804. These, based on technology he had copied in France, were sufficient to impede Boulton & Watt's quest to develop a gaslighting package. If Winsor's lectures were uninspiring – English was not his first language – the accompanying demonstrations more than compensated. He had started in 1804, at the Lyceum Theatre, dramatically lighting an ornate chandelier. Though Winsor's spectacular displays of gaslighting on the walls of Carlton House and in Pall Mall left Chalmers unmoved, their purpose was to attract investors into a company selling gas as a public utility from a central source. While his 'volley of colourful pamphlets and advertisements' has been described as absurd, Winsor had imagined the future of gas supply, with a vision eluding that of James Watt.[60]

Winsor, then, while not convincing the hard-headed Scot, was 'a persuasive entrepreneur with a great flair for promoting extravagant projects'.[61] The shows really were public, but with commercial aims, so any education and enlightenment was incidental to that end. Here was a science showman, rather than a showman scientist. The basis of patronage had shifted between the seventeenth and nineteenth centuries, and so too had the 'public' whose sponsorship was sought.[62]

Banks and Winsor exemplify two rather different ways in which science was presented to the public in their time, around 1800. For both, it was a matter of business, though Banks served a higher ideal. How, though, do these models of lecturer compare with what was around a century earlier? The starting point is a kind of outreach from the Royal Society, in the absence of such initiative from the society itself, by technicians and philosophical demonstrators who found a milieu in coffee houses and other less formal settings.[63] These were disciples of Newton, eager to spread his principles. They tended to be university-educated, working alongside an instrument maker using apparatus designed specifically for demonstration. This practice fanned out from London into the provinces, as demand for lectures grew. Over time, lecturers started to publish their talks, alongside images sufficiently detailed that the equipment could be replicated.[64]

The century's most popular lecturers were gifted in presentation. Several started as assistants to famed pioneers of science, experimenting and assisting public displays alongside the great man. Joseph Priestley, who reckoned demonstration of greater value than written accounts of experiments, was assisted

---

60 Clow and Clow, *Chemical Revolution*, 417–18, 433–4; Williams, 'Frederick Albert Winsor (1763–1830), gas engineer', DNB.
61 Williams, 'Frederick Albert Winsor', DNB.
62 See Biagioli, *Galileo, Courtier*. For this, thanks to Barbara Hahn.
63 Ashworth, *Industrial Revolution*, 154–7, extensively quotes Larry Stewart. Wootton, *Invention of Science*, 475, lists some of these followers of Newton.
64 Morrison-Low, *Making Scientific Instruments*, 268–9; see above, pp. 88–9.

Plate 9. Interior of an Alchemical Laboratory, mid-eighteenth century. It has been speculated that this is William Lewis's laboratory, or perhaps that of Ambrose Hanckewitz, assistant to Robert Boyle. Source: Wellcome Library, b21436824.

by John Warltire and Adam Walker (1730–1821). James Ferguson (1710–76) worked closely with Erasmus Darwin and others. Walker's later astronomical spectaculars, staged in provincial theatres from the 1780s, attracted audiences of up to 800, placing him in an emerging category of showman scientist.[65]

These former assistants equated to the 'technician' class. While close connection to a celebrated man of science might pave a way for life as a travelling lecturer, of course not all chose that path. Alexander Chisholm, William Lewis's devoted aide, joined Josiah Wedgwood as his scientific adviser after Lewis died.[66] John Dalton received assistance from William Henry, particularly experimentally, but Henry pursued his own career and research agenda. There are gradations here: active experimentalists who published and taught as part of their main activity, working with assistants; and the itinerant lecturers, of varying types. Lewis and Dalton arranged their lives to minimize travel and teaching commitments, in order to concentrate on laboratory work and other investigation. They were firmly of the first category, and certainly not engaged

---

65 Bruton, 'Shropshire Enlightenment', ch. 6, esp. 217, 228, 230–4; H.S. Torrens, 'John Warltire (1725/6–1810, public lecturer', DNB.
66 See pp. 77–81.

with lecturing circuits.⁶⁷ In fact, the number of peripatetic lecturers was tiny. Desaguliers believed that Europe had just thirteen such in the 1740s, of whom James Stirling and himself were the only ones interested in English industry.⁶⁸

On Desaguliers, opinion is divided. Mokyr sees him as an engineer and mathematician who embodied the 'industrial enlightenment' and bridged a 'shrinking gap between science and engineering'.⁶⁹ This is contested, not least the idea of an 'industrial enlightenment', with its subtle inference of scientific insights bestowed on industry from on high. Desaguliers is most fascinating for his role as lightning rod for a range of historical conclusions.⁷⁰

Desaguliers was connected into science at the highest level. Like Warltire and others later, he came to notice as an experimental assistant, though in his case to none other than Isaac Newton. Desaguliers was fortunately situated to take that position when Francis Hauksbee died in 1713, and then, in 1716, also followed Hauksbee as Royal Society curator of experiments. Desaguliers remained London-based, though periodically lecturing and demonstrating in the provinces and continent. He was a skilled instrument maker and self-promoter. While 'adept at cultivating aristocratic patronage', he was never, it seems, well-off.⁷¹

On behalf of his patron the Duke of Chandos, who had interests in soap-making, ore-refining, and mining, Desaguliers investigated industrial problems.⁷² This kind of activity was, however, a small portion of Desaguliers' work. His dozens, possibly into the hundreds, of papers published in the Royal Society's *Philosophical Transactions* embraced an astonishing range, from natural phenomena and physics to instrument design and contemporary agriculture and industry. Bruton observes that Desaguliers aimed to entertain as well as educate, and pursued knowledge for its own sake.⁷³ The focus was Newton's legacy, but the breadth of his curiosity is to be respected. So too is Cardwell's view of Desaguliers' *Course of Experimental Philosophy*, published in two volumes, 1734–44. In the field of mechanical sciences, says Cardwell, the work was 'magnificent', of a standard which John Banks and others, writing decades later, would not match.⁷⁴

---

67 Describing Lewis as an itinerant lecturer, and implying the same of Dalton, are among the casual misconceptions in Jacob, 'Mechanical Science on the Factory Floor'.
68 Ashworth, *Industrial Revolution*, 156.
69 J. Mokyr, featured review of Cookson, *Age of Machinery*, *Jnl Econ. Hist.*, 78/4 (2018), 1252–7, esp. 1254.
70 Musson and Robinson's poorly analysed account of Desaguliers has been widely (and often uncritically) cited: *Science and Technology*, 33–49.
71 Fara, 'John Theophilus Desaguliers (1683–1744), natural philosopher and engineer', DNB.
72 Fara, 'Desaguliers', DNB.
73 Bruton, 'Shropshire Enlightenment', 281.
74 Cardwell, *Organisation of Science*, 21 fn.

However – and it is a substantial 'however' – none of this makes Desaguliers an engineer, or an industrial innovator on any scale. He was heavily engaged with the contrasting demands of paid commissions, and on his own quest for knowledge as Newton's standard-bearer. His publications show a man intent on measuring the earth and understanding electricity – and much more – while also addressing practical solutions on themes like lenses for eyesight problems, or water-flow management in pipes. These are removed from the era's significant long-standing industrial challenges. Overwhelmingly, the evidence of how real progress was made in such instances – think of puddling iron, generating gas, spinning cotton by machine – points to three elements: solutions resting on a mix of engineering skill, industrial knowledge, and dogged determination. Remember Watt's observation that the most difficult part was 'the making of it usefull'.[75]

Desaguliers did attend to practical matters. Yet his impact – and that of other public lecturers of his generation – in achieving outcomes of real value to industry, is not proven. What, for example, is to be made of his account of seven steam engines he 'caused ... to be erected', the first for Peter the Great in 1717–18? Desaguliers' own role is opaque, couched in vague terms, a matter for conjecture.[76]

The world of the cosmopolitan Desaguliers was polite science. Did he in any direct and useful way engage with the rising provincial industries? There are two episodes in his career which define – much of this in his own words – the limits of his influence in key sectors. The first is Desaguliers' very promising finding on the economy of waterwheels. Emerging from a study of corn-mills, the discovery included calculations on waterwheel efficiency which presaged those of Smeaton in the 1750s. Although Desaguliers' experiments cannot have achieved the same degree of accuracy and certainty which Smeaton's mathematical modelling later enabled, he anticipated Smeaton's conclusion in favour of overshot wheels.[77] This could be considered a breakthrough, yet it had no detectable impact.

Contrast this with Smeaton, whose site-specific templates for water power were based upon neat and incontestable calculation. He produced the models in full knowledge that they addressed an urgent need in northern England and would speedily translate into improvement on the ground. With the best watermill locations fully occupied, projectors must increasingly fall back on awkward sites in ever more problematic places. To Smeaton, disseminating useful knowledge was the point. The matter was clear, that industrially

---

75 Univ. of Glasgow, Archives & Special Coll., GB 247 MS Gen 501/27, letter from James Watt to James McGrigor, Glasgow, 30 Oct. 1784.
76 Desaguliers, *Course of Experimental Philosophy*, II, section XIII, 488–9.
77 Cossons (ed.), *Rees's Manufacturing Industry*, III, 477–8; see p. 86 above.

dynamic districts local to him must wring the last ounce of power from available water sources, and he addressed it.

To a modern reader, Desaguliers' *Course of Experimental Philosophy* appears largely a work of theory, presenting his lectures from c. 1713 on the laws of mechanics and aspects of the physical world, accompanied by less substantial technical insertions.[78] Theory, however, was not a word that Desaguliers much used. When he did, the meaning was not as now understood. He distinguished between 'mechanical principles' and 'the theory of engines'.[79] This is explained in his observations on waterwheels. 'Engines' referred to 'hydrostatical and hydraulick machines'. While he did separate theory from practice, 'theory' to him meant a design or a plan, something altogether more concrete than a modern understanding of theory. Much else of what he wrote was a critical commentary on recent developments. He approved Hadley and Sorocold, 'the only engine-makers of any note of the last Age of England'.[80] But otherwise Desaguliers disliked much of what he saw.

His targets included 'quacks' ignorant of mechanics; pure mathematicians who emerged as inept project-managers; and workmen prejudiced against academics. He condemned unsuitable designs and poor execution, and much more. Sorocold and Hadley succeeded because of their 'strong natural genius for mechanicks'. Without knowing mathematics or philosophy, they 'performed great works' based on their observations, a 'natural mathematics of their own'. Waterworks, continued Desaguliers, were complex.

> He that would meddle with Water-Works, should know so much of Mathematics, as to understand mechanical Principles; be so much a Philosopher, as to be skilled in Hydrostatics and Pneumatics; and be so good a practical Mechanick, as to know the Nature of Materials, and how to put them together in the best manner.

Desaguliers also took aim at projectors in waterworks whose priorities were profit and acclaim, rather than serving the needs of clients, who were thus drawn into impractical schemes. 'Now', he wrote, 'we are overrun with Engineers (not *Ingenieurs* of a proper Education for the Science)'.

---

78 For background, see Anstey and Vanzo, 'Early Modern Experimental Philosophy'.
79 Desaguliers, *Course of Experimental Philosophy*, II, 35.
80 Chrimes suggests a tantalizing possibility – that Sorocold trained under Cornelius Vermuyden II, of the Dutch family of engineers famed for draining the East Anglian fens: Chrimes, 'George Sorocold, 1668–fl. 1716', BDCE. If so, Sorocold's importance may have been understated. He and Hadley worked on many water-supply schemes across England, using water wheels to pump river water into towns. John Hadley (not to be confused with a later mathematician and instrument maker of that name) patented a device for raising and lowering a water wheel in 1693 and was surveyor and engineer of the Aire and Calder Navigation, 1698–c.1701: A.W. Skempton, 'John Hadley (fl. 1689–1701)', BDCE.

Almost all the Plumbers and Mill-Wrights now set up for Engineers, tho' I hardly believe there are two of them who know how to measure the quantity of water required to turn an Undershot, or an Overshot, or a Breast-Mill. They only judge of a Stream by the Eye, and he that has most practice is likely to succeed the best.

As for pure mathematicians, compared with 'mixed' mathematicians they were deficient in understanding 'mechanical performances'. This placed them at the mercy of 'the ignorance or knavery of workmen' when undertaking waterworks, says Desaguliers. The theory, the design, would fail, because (in essence) the workmen decide that it cannot work, and the designer is unable to produce a sufficiently detailed scheme to convince the workmen of his case. For 'a compleat Theory', the undertaker must understand the work of bricklayers, masons, millwrights, smiths, and carpenters, as well as the underlying principles of water engineering.[81]

'*Ingenieurs* of a proper Education for the Science' is illuminating, and so is Desaguliers' support of Sorocold and Hadley, putative successors of the seventeenth-century Dutch fen-drainers. He suggests the pair's 'strong natural genius for mechanicks' forgave any scientific shortcomings. There is no doubt they were highly competent civil engineers. But to Desaguliers, they were not *ingenieurs* in possession of a 'compleat theory'. By implication, that was how Desaguliers saw himself. Yet, while he had insight into large technical projects, and some of his criticisms appear sound, he was not an engineer as we would understand it. His 'theory', his scheme, of waterworks, if it found application at all, was not executed by Desaguliers himself.[82]

The second episode revealing something of Desaguliers' approach was about steam power – fire-engines, as he knew them.[83] James Watt, introduced at the University of Glasgow to a Newcomen engine alongside a working model of the same, sought to explain why the model was far less efficient. There was little useful guidance to be found, so Watt turned to Desaguliers' published works. Quickly he concluded that the philosopher had not grasped steam's mysteries. Nor at that stage had Watt, but he made practical progress, through his own observations and experience of materials gained as an instrument maker. The science of latent heat would catch up later, through his scientist friend Joseph Black.[84] Watt was learning the lesson of 'the making of it usefull'.

---

81 Desaguliers, *Course of Experimental Philosophy*, II, Lecture XII, esp. 412–16.
82 Desaguliers, *Course of Experimental Philosophy*, II, Lecture XII, 414–15. For changing definitions of 'engineer', see Cookson, *Age of Machinery*, 76–8.
83 Desaguliers, *Course of Experimental Philosophy*, II, section XIII, 464–90, on fire engines; 532–4 for a postscript on Newcomen.
84 Marsden, *Watt's Perfect Engine*, 31–2; Ashworth, *Industrial Revolution*, 175.

There is more to this than a mere misunderstanding by Desaguliers. Some of his own words show that he allowed conceit to override any veneer of objectivity. He was particularly dismissive of, and very wrong about, Thomas Newcomen.

> Desaguliers' writing ... reveals the gulf that then separated the scientist from the practical craftsman, for it betrays a complete ignorance of the working principle of Newcomen's engine and a failure to understand the *modus operandi* of its ingenious valve gear.[85]

That is Rolt, an authority in matters mechanical. Desaguliers' ignorance and prejudice, says Rolt, tainted Newcomen's legacy, and hence the engine, which was in fact 'a most potent agent' in 'quickening the pulse of industrial revolution'. Desaguliers could not bring himself to believe that two tradesmen from Dartmouth – Thomas Newcomen, iron merchant, and John Calley, a plumber – could achieve such a thing. His version was that Newcomen picked up the technology from Savery, who had been unable to apply his engine to draining mines. After 'several experiments', 1710–11, Newcomen and Calley tried their engine in the Midlands:

> after a great many laborious attempts, they did make the engine work; but not being either philosophers to understand the reasons, or mathematicians enough to calculate the powers, and to proportion the parts, very luckily by accident found what they sought for.

And then, says Desaguliers, first Birmingham workmen improved the pumps; then a method of condensing within the cylinder happened through chance; and subsequently others variously improved the design.[86]

That narrative can be disproved. So can claims, one by Joseph Black, that Newcomen must have known about Papin's concept of an atmospheric engine, which had been revealed to the Royal Society in 1675. Rolt insists that Newcomen had no contact with London's 'closed shop' of savants and instrument-makers, and this is backed by contemporary evidence, which also undermines ideas that Newcomen's innovation derived from Savery. The Swede Mårten Triewald, who met Newcomen and Calley in 1716, was convinced they had no prior knowledge of Savery, and that Newcomen 'for ten consecutive years ... worked at this fire-machine'. Newcomen was pursuing ways to solve drainage problems in Cornish tin mines, which he visited on business. His engine was trialled near Dudley Castle, in a district he knew as a buyer of iron, and one where he could source the parts he needed. That the atmosphere had weight had been known for a long time; and so too the potential of vacuums, which Papin created experimentally.

---

85 Rolt, *The Mechanicals*, 5.
86 Desaguliers, *Course of Experimental Philosophy*, II, 466–7, 532–3.

But, says Rolt, contemporary scientists insisted that Newcomen must have seen Papin's plan – which was never made to work – and 'refused to accept the simple truth that [Newcomen's engine] was the outcome of over ten years of patient experiment, of trial and error, by a country craftsman'. They played down this landmark achievement, an engine so fine that it remained little altered in fifty years, 'because they could not bring themselves to believe that such a man had invented such a machine'.[87]

Desaguliers was not an engineer, and nor was he a fair-minded arbiter on Newcomen. That does not cheapen his undoubted achievements. Rather, it confirms the environment in which he worked, and some prejudices reflecting his time and class. Such views continued into William Murdock's time, when his gaslighting advances earned the Royal Society's Rumford medal in 1808 – except that he was considered unsuited to present his own paper. It was written by the Watts and William Henry, and delivered by Sir Joseph Banks. Worse, the honour was pursued, not as an accolade for a loyal and talented employee, but with cynical intent. Because the confusion of prior claims ruled out patenting, Boulton & Watt saw this public acknowledgement as a means of establishing their firm's priority in gas-plant manufacture.[88] So Watt's most innovative employee, of a highly respectable family (acquainted with Boswell, who possibly introduced him to Watt), was marginalized because he came from the class of millwrights and land agents. He was not of appropriate status.[89] This suggests the Lunar men's view of themselves. Élitism was alive and well, here dividing an evidently close-knit group of innovating technologists working in the same business. Class cut through collaborative endeavour. Even within science itself, the new term 'scientist' was unbearable for those lofty philosophers who would not consider themselves equivalent to men in receipt of salaries.[90]

---

87 Rolt, *The Mechanicals*, 4–5; Allen, 'Thomas Newcomen (1664–1729), ironmonger and inventor of the atmospheric steam-engine', DNB; Cossons, *Industrial Archaeology*, 58–62; Ashworth, *Industrial Revolution*, 173–5. The first Newcomen engine supplied to a Cornish tin mine was in 1710, the last in 1778. The second, in Dudley, served a colliery; the last (of more than 1,200 Newcomen engines in Britain) was built *c.* 1803, in Gloucestershire: J. Kanefsky, Early Engine Database, https://coalpitheath.org.uk/engines/. For a view at odds with Rolt, see Wootton, *Invention of Science*, 507–8. Wootton, generally admirable, does not challenge Desaguliers' account sufficiently, I think, and overlooks the significance of contexts.
88 Clow and Clow, *Chemical Revolution*, 432; Griffiths, 'William Murdock', DNB.
89 Griffiths, 'William Murdock'.
90 Ross, 'Scientist', 65–6.

Posterity condescending

While industrialization is a story of dynamic change, it is also one of settled practice and, for most, that meant inhabiting a context determined by social status. These contexts did not, however, stand still. The taste for innovation was not limited by class. Ideas appeared from unexpected directions, and in different guises.

Yet it seems that the originality of lowly individuals was consistently underrated. The exception is where such individuals were the subject of a rags-to-riches narrative. 'Lowly' extends to some who had bettered themselves and moved into 'intermediate' occupations, but never quite fitted there because of how they presented themselves. Men more obviously respectable were less doubted, and their achievements better acknowledged. Victorian writers' prejudices still play out, still insufficiently challenged. So we too condescend, about people of intelligence and vision who achieved things which most of us, with all our present-day advantages, could not.

The layers and nuances of social status are one reason why it was important to define a millwright. Within that one occupation existed many strands, different activities, and social gradations.[91] In clarifying the differences between a millwright, of whatever rank, and smiths and ironworkers, those trades' contributions to new kinds of engineering become more visible.

The less-resourced innovators also faced greater difficulties in protecting and profiting from their discoveries, and often lost control. This was Samuel Crompton's problem. His spinning mule vies with Newcomen's engine as the century's greatest breakthrough, combining the best features of Hargreaves and Arkwright machines into a spectacular hybrid in 1778–9. Crompton set out to solve a specific problem, and the result was beyond any doubt his own work – designed and constructed by an impoverished 21-year-old limited in metal-working skills, though mathematically educated. He took no steps to protect the technology, which spread rapidly and was the foundation of modern cotton production. Crompton's epic achievement brought him very little financial reward.[92] He made his name but lived out his life in poverty.

This contrasts starkly with, for instance, the care taken by James Watt in guarding his own interests. The famed innovator actively impeded others from enhancing his steam engine. This was more than patent protection. In blocking his employee Murdock's further improvements to the patent engine, Watt exposes the firm's tactics. Defending the monopoly was preferred to upgrading their own technology. Boulton & Watt's reputation and access to

---

91 See ch. 4, 'Explaining millwrights'.
92 D.A. Farnie, 'Samuel Crompton, 1753–1827', *DNB*; Hahn, *Technology in the Industrial Revolution*, 81–5; Catling, *Spinning Mule*, 31–40.

finance opened up strategies denied to Crompton, Hargreaves, and Cort, and others with few means.

Was Watt further advantaged through his privileged access to natural philosophy circles? Not directly by the Lunar Society connection, it seems, though that (and his wider reputation) eased him into the Manchester Lit. & Phil. There the worlds of science and business were unusually, perhaps uniquely for the time, enmeshed. While the revered and ascetic Dalton, prioritizing higher truths over worldly riches, was the society's nucleus, he was atypical. Within the Lit. & Phil. a strong and informed interest in science coincided – especially in bleaching and dyestuffs, and developing coal-gas technologies – with some leading members' business pursuits. It is no coincidence that this society thrived in that great booming cotton-manufacturing town. The two James Watts sustained a close relationship with the Lit. & Phil., the younger serving it with distinction. Ultimately, they could not make the connection work to their advantage. The outcomes did not fit with how they did business.

George Augustus Lee and his associates better illustrate the narrowing gap between scientific principles and new technologies. Lee's connections, some very eminent, joined his mission to work out new industrial possibilities from ideas not yet fully packaged. Around 1800 they broke new ground, notably through the gaslighting experiments in his own factories.

The Lit. & Phil., a valuable meeting point and source of specialist support, drew heavily on wider networks. Its facilities would have advantaged any developing industrial centre, but there is no evidence of anything quite like it elsewhere, among the generality of Lit. & Phils. The Manchester group is best measured against the context of go-ahead industrializing communities, where boundaries were shifting or fading. Philosophers were drawn into industrial opportunities, and industrialists explored the promise of scientific discovery.

Here is the wider picture – across regions and industries, science and technology ebbing and flowing in their relative significance. The most pressing questions changed. The generalizations of 1750 may not be valid in 1800. Local environments could be decisive, and so could bigger contexts. It was possible, of course, that people, as well as technologies or theory, could be wrong before being right, or be selectively correct, or after early promise turn out never to have been right at all. Theory ran ahead of mechanical practice, though it was barely accessible to those most needing it; by contrast, chemical practice advanced theories which came to underpin chemistry. And just as certain early textile engineers quickly realized that they could make more money more easily by using the machines rather than making them, so the business side drew in people whose first interest had rested in science. In all of this, there is no ignoring the significance of networks and localities.

6

The shape of networking

Networks and localities were key to how innovation was supported and sustained, and to how industrial practice evolved. And eighteenth-century industrial organization – in all its manifestations, large and small – was itself underpinned and informed by networking.

Success did not follow a single route. Things could be, and were, done differently, on different shop floors and in different regions. Promising advances were disseminated through inter-regional connections, active and significant even quite early in the century. Industrialization rested on more than one viable model, most strikingly illustrated in Scotland's individualistic course. There, an idiosyncratic path was laid by economics, politics, and culture, and born out of constitutional crisis. Public policy for once responded: its catalyst was the Board of Manufactures. Picked up by academics, increasingly falling into alignment with Scottish university interests, the Board supported an industrial renaissance north of the border. Nothing comparable is found elsewhere in Britain. The Scots contribution to industrialization has not been justly appreciated.[1]

Within England and Wales, regional interconnections map the shifting scenes of innovation and development. Those relationships are descriptive rather than quantifiable, and they colour and deepen the usually dry fare of business papers. Tracking human associations, and relating them to industrial activity, gives substance to the gathering momentum. What follows here is an exploration of the different forms connection took: industrial organization, access to finance, infrastructure, combines, communities, migration, and, ever present, the shape of networking.

Regions

Everywhere needed iron. New markets were building in iron products for industrial and domestic use, notably in the Atlantic trade. Yet a national market in iron products struggled to emerge in Britain because of the poor state of transport links. This accounts for the Darbys' signature cooking pots selling in

---

1 The Scottish experience is detailed in the following chapter.

vast numbers locally in the west Midlands, and as exports to North America, while trade with the rest of England was much less extensive. The natural route for goods from Shropshire was via the River Severn and by sea from Bristol. The Black Country's poor connections to north and eastern England had determined Ambrose Crowley's move to Newcastle. The relative isolation of Shropshire ironmaking – though it was an important centre of skill and innovation – also accounts for its own distinctive character.[2] But it is also true that regional interconnections between iron producers were sufficiently close to shape new technologies and practices, even before 1700. People moved more easily than bulky iron products.

In this century of wars, securing supplies of part-processed Baltic iron was a national strategic necessity. Still, the direction taken by the British iron trade, in matters of investment and infrastructure, was left to private choices. Unlike Sweden and Russia, Britain had no national body taking an informed interest in iron production. Even in Scotland, the Board of Manufactures' eventual success – across a range of industries – rested heavily on the initiative of a handful of influential individuals.

The great ventures which brought regions together in tangible form were instigated locally, promoted by merchants and landowners. A navigable waterway, fabricated from rivers or freshly channelled, connected towns and ports. It could change the direction of traffic, and established new bonds from the time that people united (not always entirely amicably) to plan and invest. But there it was, a palpable link between regions, a joint enterprise, an infrastructure supporting many industries and enabling more such projects to follow. If evidence is needed of the value of a canal, consider a locality without one. Tees valley coal and textile interests made strenuous but doomed attempts to find a plan that was geographically and economically viable, to link Darlington and its hinterland with the North Sea. Despite taking the best advice available – from James Brindley and Robert Whitworth in the late 1760s, and John Rennie in 1812 – there was no success. The region remained a relative backwater. Until, that is, attention turned to building a railroad, with famous results. The Stockton and Darlington railway opened in 1825.[3] Nor should the tardy arrival of the Somerset Coal Canal be forgotten, transforming the Wiltshire textile trade a little too late.[4]

A timely waterway could be transformative. The Severn, though prone to flood and in places hard to bridge, in connecting the west Midlands to Bristol and the sea, was fundamental to the growth of Shropshire's iron trade. This was some while before a canal network began to connect northwards from there, to

---

2 Hayman, *Ironmaking*, 32-3; 'Shropshire wrought-iron industry', 42–6, 214–16.
3 Cookson (ed.), *Victoria County History of Co. Durham*, IV, 127–33.
4 See p. 17.

ironworks in Cheshire and north Wales.[5] Waterways opened many and varied possibilities, with coal haulage, inward or outward, the outstanding benefit. There were obvious advantages for other bulky or heavy goods, materials and products, and fragile items like Wedgwood's pottery.

The impact of the Aire and Calder navigation on Leeds, the cheaper coal and easier access to external cloth markets which it enabled, has been discussed.[6] But when Angerstein passed through in 1754–5 on his way from Barnsley to York, he paid Leeds scant attention. He remarked that it was the country's main centre for the woollen trade, and mentioned two steam engines at work in coal quarries nearby.[7] Much of the industrial activity feeding the cloth trade happened out of town, and Leeds itself had little to interest Angerstein. Kirkstall Forge was then in the doldrums, approaching the end of the Spencer era. He saw no novelty, no iron-processing worth noting. Twenty years later, a visitor would see the Yorkshire stages of the Leeds and Liverpool Canal coming into being, passing close to Kirkstall Forge and setting it up for a new era when the east and west coasts connected. But there were immediate benefits for industry. From the later 1780s, access via a branch of the Leeds and Liverpool to coal and iron reserves south of Bradford enabled other ventures in mining and ironmaking. Those in turn presented new options for Leeds.

The benefits to textiles were less obvious, but very real. Water transport was not a conduit linking stages of production, as might happen in other trades. During the domestic era, textile processes generally connected by road, with weary agents collecting and delivering to cottages and small workshops by cart or packhorse. But increasingly the raw materials – flax or cotton, and even wool – were brought into the region by water. Coal, in growing demand to fuel steam-driven machinery and textile processes such as dyeing, was carried by boat wherever that was possible. The only direct encounter of processed textiles with water transport may have been when finished cloth left textile towns for Hull, destined for export.[8]

Canals were multi-purpose, then, and adapted to new functions beyond anything that their projectors could have foreseen. The inter-regional benefits were obviously substantial, the cross-locality links important. For an industry like machine-making in Leeds (and Keighley, Manchester, and other engineering centres), having waterways to transport heavy parts and materials inwards,

---

5   No bridges crossed the Severn in the vicinity of Coalbrookdale before the famous iron bridge (1777–80), and the short-lived first wooden bridge (1780) at Coalport: Newman and Pevsner, *Shropshire*, 635, 638–40.
6   See pp. 10–11, 22.
7   Berg and Berg (eds), *Angerstein's Illustrated Travel Diary*, 221.
8   For the Yorkshire textile industry's geographical spread, see Giles and Goodall, *Yorkshire Textile Mills*, 4–5, Fig. 2; Heaton, *Yorkshire Woollen and Worsted Industry*, 287, and *passim*; also ch. 1, above.

and finished machinery out, was of clear advantage. But it is not all. The new engineering did not grow and innovate merely by moving things around. The canal is to be seen in a bigger context of human connection, another supportative technology. The personal links formed by waterways, in planning and using them, and establishing new trading contacts, were of significance.

London has been acknowledged as a centre of markets and finance, a place to train and improve skills, or as a source of enlightenment and scientific thinking. The capital's more oblique role as a place of introduction and an information exchange should also be noticed. Home to one in ten of the British population, London was always a great magnet: for employment seekers; for merchants buying or selling, for instance, iron or textiles; and for the generally inquisitive, interested in progressing their interests. It was a place of casual encounter, a meeting of regions and nations. Here occurred significant exchanges of intelligence, random as well as specific. Data could be valuable: it was noted, reported back to acquaintances at home, and noted again by keepers of memorandum books, such as John Brearley.[9] The traffic in relatively casual information snippets is not measurable, and nor are its effects. But these habits were entrenched by the last quarter of the eighteenth century, and they had value for innovation as well as commerce.

Networks and migration

Communities and networks were not static. People moved, taking with them ideas and technologies. Building new connections was a constant process, between regions and also internationally. From the class which kept the best records – business owners, managers, professionals, and similar – comes much evidence of interaction. Less accessible are exchanges between skilled manual workers, those adaptable shop-floor specialists increasingly sought as employees or subcontractors in key industries. A few are documented through their dealings with larger businesses and projects. Less apparent is the nature of relationships these men had with each other, through their own small businesses or as co-workers in part of a larger enterprise. These contacts matter, for workers' connections and movements were fundamental to how innovation happened.

As a group, skilled iron workers, smiths, and millwrights were on the move, deployed to sites away from their base or relocating to new positions. Informed and encouraged by associates and contacts, their willingness to move supported the advance of key industries. Such movements, whether permanent or stages in a career, had qualitative significance. It was not just about numbers, about

---

9 For Brearley, see pp. 15–16.

increasing the pool of key workers, though that was important. The switches did not eliminate variations in regional ironworking, but in bringing together workers from different ironworking traditions they fostered an atmosphere of inquiry and openness to innovation, as Evans describes in the south Wales workforce. Remarkably, Cyfarthfa ironworkers had their own philosophical society from 1807, some way ahead of polite circles in Shropshire, the county of some of their forebears.[10]

It was curiosity about a particularly significant diaspora, the ironworkers migrating northwards, from Wortley and other south Yorkshire furnace and forge districts to Leeds and Keighley and thereabouts, which sparked this research. The impact on west Yorkshire machine-making was monumental. The migrants gave a foundation for new ways of working, and the technical capacity to create a leading centre of engineering after 1800. That much is known.[11] Less clear are the 'push' factors, and the information networks promoting west Yorkshire's attractions. And also, what did migrants bring and disseminate, in the shape of techniques, processes, and organizational models?

By their nature, because skill and industrial knowledge are learned and stored and rest within someone's memory, as assets they are beyond computation. But there is a physical presence, a person who arrives because their expertise is in demand. They bring the potential to spread by example, to reveal their specialism to co-workers and beyond. During the sixteenth and seventeenth centuries, in every sphere of industry, Britain had drawn heavily on foreign proficiencies, particularly from continental Europe. Sometimes explicitly encouraged by the state, and on other occasions imported by religious refugees settling permanently, new aptitudes were introduced to staple trades. Dutch engineers drained the fens, a domestic silk industry grew on French Huguenot skills, and workers from across Europe helped develop glass-making, brewing, and many other trades.[12] Abraham Darby is said to have brought Dutch foundrymen to Bristol in 1704 to assist with his pot-making trials, and would later transfer his valued workers to the new business in Coalbrookdale.[13]

Introducing foreign competencies was significant, but movements were not one-directional. Crowley's company provided housing as an inducement to attract skilled workers to the north-east, from London and other iron-producing regions. But as British talent became more sought after, even at that early date came efforts to entice managers and other ironworkers to Russia, and later to Sweden. The first legislation to counter this passed early in the reign of George I.[14] Whatever the ups and downs of continental relationships during this century

10 Evans, *Labyrinth of Flames*, 75–6, 205.
11 Cookson, *Age of Machinery*, esp. ch. 6.
12 Ashworth, *Industrial Revolution*, esp. 85–91, 146–54.
13 Ashton, *Iron and Steel*, 27, 197–205.
14 Ashton, *Iron and Steel*, 197–205.

of war, international links within the iron trade seemed only to flourish. This continued into the era of mechanical engineering, when Matthew Murray and associates in Leeds built on Carron's links with the Russian court, while also establishing a strong relationship with Sweden. For Murray this was about selling steam engines, but it involved an invigorating exchange of personnel.[15]

Water transport improvements reinforced south Yorkshire's external links. There, a mid-century boom coincided with the Don's upgrading, to the North Sea from Rotherham (1739–40) and then Sheffield (1751).[16] This eased the passage of high-quality iron ore from Sweden into Sheffield and its surroundings. From the late 1740s, the Cutlers' Company, the regulator of Sheffield's cutlery trade, proactively encouraged skilled inward migration. Advertising in London and other regions, they challenged any impression that outsiders were unwelcome. Sheffield offered good coals and water, the best grinding wheels in the kingdom, and merchants ready to assist with marketing and other practicalities, the company suggested; nor would arrivals encounter problems from churchwardens about their legal settlement. The navigation also encouraged new iron and steel technologies, and more ambitious projects. On the Don at Masbrough, the Walkers invested in substantial new plant from 1746. In Sheffield, Benjamin Huntsman, possibly himself of German or Dutch descent, was producing crucible ('cast') steel by 1750, the innovation soon impacting upon local steel output.[17]

Why, then, with Sheffield apparently booming, would so many workers with marketable skills choose to leave the region? Sheffield industries and the local iron trade around Wortley renewed and prospered from the 1740s. In the longer term, however, their paths differed. The Wortley parishes' staple trade, nail-making, hit hard times, and much of the displaced labour could not be accommodated in Sheffield or in local iron production.

Sheffield was physically constrained. As noted, by 1780 no vacant space for water-powered manufacture remained within the town.[18] Its exceptional grinding facilities accounted for the superiority of Sheffield cutlery, with riverside sites rented out by the hour. Demand for premises came, too, from trades auxiliary to cutlery – handles, boxes, silver plating – and related products, fine items like surgeons' instruments. The workshop sector was diverse, and

15 Cookson, *Age of Machinery*, 276–9.
16 Tankersley iron not being of satisfactory quality for Sheffield cutlery, from the sixteenth century Spanish iron was imported, as were artisans from France and Sussex. Before the Don was reliably navigable, Hallamshire imports and exports moved via the port of Bawtry and outwards via the Trent and Humber: Hey, *Fiery Blades*, 55, 300; and see Riden, 'Navigation on the Don before 1726'.
17 Hey, *Fiery Blades*, 165–6, 183–4, 306; 'Benjamin Huntsman (1704–76)', DNB; Hunter, *Hallamshire: A New Edition*, 170–1; see below, pp. 161–5, for the Walker Family.
18 See p. 87.

highly specialized. There were also metal-ware trades less directly connected to cutlery: the manufacture of screws, spindles and flyers, files, wire, and hand tools of various types, specialities later highly valued by textile engineers.[19]

In contrast, the small workshop sector around Wortley was struggling. The parishes of Tankersley, Silkstone, and Penistone – their boundaries met on the Don at Wortley Upper Forge – and especially Ecclesfield, south of Tankersley, were nail-making centres and had been important users of Wortley iron. Their decline shows in the outwards migration from that time, and in the continuing relationships between those who left for west Yorkshire, some of them building engineering businesses there, and family members remaining in the Wortley neighbourhood.[20]

Nail-making became a staple trade in the district after slitting mills were introduced, c. 1600. Many nailers worked in smithies at home, using locally produced rod iron, rolled flat and slitted into lengths. Their modest enterprises sat within a vast, vertically integrated, trade. The organization linked iron-mining and iron-processing, to slitting, to nail-making and merchanting. Ironworks clerks communicated directly with chapmen who delivered rod iron to nailers and co-ordinated the industry.[21] The Spencer combines, dominant in south Yorkshire iron, managed at least five slitting mills among their Yorkshire ironworks in the 1730s. Slitting and rolling mills continued under Spencer successors at Wortley, Masbrough, and other sites, including Kirkstall and Colne Bridge. There were slitting mills alongside wire mills at Thurgoland, and also at Seamer Forge.[22]

Nail-makers' pay was dire, driven down to the level of unskilled labour. It was usually supplemented by an agricultural holding, with forge work laid aside to accommodate farming's seasonal demands. The apprenticeship, formally seven years, could probably have been done in two, but was deployed to restrict entry to the trade, for even those poor wages were under threat. In the 1730s, female and child nail-makers in the Midlands undercut them. In response, local chapmen beat down the price of rod iron supplied from Wortley and Rotherham forges, securing nail-making's immediate future in south Yorkshire.[23] But the nail-making economy remained fragile.

19   Tweedale, *Steel City*, 42–3; Berg, *Age of Manufactures*, 263–4; Hey, *Fiery Blades*, 114–36, (for rivers) 179–83; also Crossley (ed.), *Water Power on the Sheffield Rivers*.
20   Cookson, 'Wortley Forge', esp. 71–2.
21   Hey, *Fiery Blades*, 93–7, 305; *Guide to Hoylandswaine Nail Forge* (pamphlet, South Yorkshire Trades Historical Trust, 2010); Cookson, *Age of Machinery*, 130–2; Hopkinson, 'Development of lead mining', 200.
22   Raistrick and Allen, 'South Yorkshire Ironmasters', 168–9; Butler, *History of Kirkstall Forge*, 5; WYAS Leeds, WYL76; Hopkinson, 'Development of lead mining', 204.
23   Hey, *Village of Ecclesfield*, 48, 55–61; Cookson, *Age of Machinery*, 130–2; Hopkinson, 'Development of lead mining', 211–12; Evans and Rydén, *Baltic Iron*, 193.

THE SHAPE OF NETWORKING

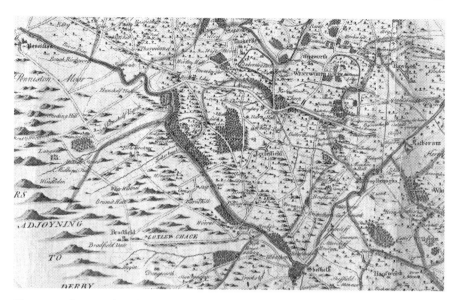

Plate 10. The Wortley district, a detail from Dickinson's plan of 1750. The River Don flowing from Penistone to Sheffield marked boundaries of the Wortley parishes. These were Penistone, Silkstone, Tankersley (into which fell Wortley chapelry, though the Wortley forges were only partly situated in that parish), and the nail-making centre of Ecclesfield. Grenoside, where the Walker ironmasters originated, is unmarked on this plan; it is west of Ecclesfield, north of Birley Edge. The site of their then new works, Masbrough, is given as Messburgh, on the lower Don next to Rotherham. Note that Dickinson's plan is orientated NNE, not North. Source: Sheffield Archives, WWM/MP/95R, Map of the Southern Part of the West Riding of the County of York by Joseph Dickinson, 1750.

William Spencer, the last active head of the family concern, left the nailing trade in 1748, after losing his Wortley sites to John Cockshutt.[24] A marked decline followed. The 1770s saw over-population and high levels of poverty stretching Poor Law capacity, particularly in Ecclesfield parish. File-cutting picked up some of the surplus workforce, as nail-making retreated. When the commons, green and open fields of Ecclesfield, were enclosed in 1789, little resistance came. Long-established residents were already drifting away, to other places and trades.[25] With nail-making much diminished and farming disrupted,

24 Hopkinson, 'Development of lead mining', 212, 218.
25 Hey, *Village of Ecclesfield*, 18, 27, 29, 55–61, 66, 119.

the district was blighted by extreme poverty. The domestic and workshop sector was hit hard, the in-demand skilled forge and furnace workforce less so.

Were nail-making techniques transferable to new occupations? Many nailers were at least semi-skilled, their pay depressed by the dearth of local options. Low pay was no measure of a worker's intelligence and capacities. In fact the nailing trade contributed substantially to the foundations of machine-making. It proved to be of far greater direct importance in that respect than occupations such as instrument-making. From the Wortley district's workshop economy, which produced many types of small forged parts, there arose a class of labour precisely suited to this era, that interlude before a mechanism could manufacture every type of metal component. In one of innovation's paradoxes, a machine world would not have been possible without handmade parts, before machines could make them.

All that would start to change after 1800, with new machine tools and specialist machinery. Before that, accuracy was achieved through manual technique, a fine judgement, and repetition. Workers in small forges, smiths and makers of nails, screws, bolts, nuts, spindles – items then coming into high demand – adapted, and were sought after. As machine concepts improved, so must the handmade parts. Precision formed by constant repetition, and to ever-higher standards – was that semi-skilled work, or something else? Whichever, the technique discovered its new market.[26]

So just as rural life became unsustainable, other options presented. The Wortley parishes' abundant smithing skills did not suit the nearest towns: Sheffield was crowded with niche trades, Barnsley had wire-drawing and high-end specialisms. Nor, in general, was the nailsmith's skillset useful to local ironworks. West Yorkshire, and especially Leeds, looked increasingly attractive, offering the prospect of joining relatives and former neighbours, engaging in work that these migrants could adjust to. It was a rational and positive response to the difficult local situation, where other options were limited or non-existent. In making that decision, adapting to change by striking out on a new path, the individual became a part – potentially a quite significant part – of the change.

Migration, whether local, between regions, or international, was largely unremarkable. It happened for a reason, yet it carried dangers, especially for those with few resources. Movement had always been a feature of the iron trade. Many instances show cross-regional traffic in skilled workers: the Carron works importing Midlands skill, Shropshire and the Black Country providing the foundations of Wortley's and Cheshire's Spencer syndicates, and (like Cumberland) feeding into south Wales. The nail-makers of the Wortley parishes, though, were more rooted, most of them also occupied as farmers.

---

26   Cookson, *Age of Machinery*, 64–8.

Nor would a large textile town necessarily hold appeal. But the relocations from Wortley to Leeds show how knowledge could mitigate risk and hesitation.

South Yorkshire's two main waves of migration to Leeds illustrate this point. The first, during the 1770s and 1780s, comprised a few skilled individuals who were sufficiently well resourced to set up in business. The second, following quickly on and intensifying after 1800, was a larger and longer movement of displaced nailers and other trades, such as carpenters, with proficiencies valuable in machine-making and iron-founding. Many were known to the first group, and had been encouraged and informed about the possibilities opening up in west Yorkshire. Familiar people at their journey's end, at Kirkstall Forge, or in the machine shops of Leeds and Keighley, helped them settle. Sheffield to Keighley was about fifty miles, the Wortley area to Leeds perhaps thirty. The arrivals were mainly young and fit, and escaping rural drudgery was matched by the thrill of a town and greater autonomy. But more than this, there was promise – to develop marketable skills in a context of galloping innovation, to flower and make their own niche.[27]

The first wave, those creating the confidence for others to follow on from south Yorkshire to Leeds and Keighley – Jubb, Gothard, Salt, and then Richard Hattersley – were matched by arrivals from other regions, notably Matthew Murray, William Carr, and (later) Peter Fairbairn. Including a good measure of indigenous talent, this group, as it assembled and grew, was instrumental in launching and developing mechanical engineering in west Yorkshire. Scots, Midlanders, and others had similar impacts upon Lancashire machine-making.[28] In all or most cases, the first arrivals were millwrights, foremen, or similar, men with specific aptitudes who had followed a post-apprenticeship 'improver' track and learned something about industrial organization.[29] Salt, Gothard, and Hattersley had honed skills and broadened their know-how in Sheffield. Hattersley became a specialist in making screws – large ones, used to fix machines and waggon-ways – which took him to Kirkstall Forge, from where in 1789 he was poached by a customer to take charge at Screw Mill, Keighley.[30] Gothard and Jubb, both ironworks millwrights, worked at, or were customers of, Wortley Forge during their early careers and afterwards extended that experience within the region. Most likely, both had some familiarity with Leeds and other west Yorkshire ironworking sites, through their employment at Wortley and Rotherham. The Walker company was a significant supplier of

---

27 Thanks to Susan Bayly for insights into twentieth-century rural-urban migrations in Vietnam and Russia.
28 Cookson, *Age of Machinery*, esp. 96–7, 100–2, and ch. 6, esp. 163–5; Cookson, 'Hunslet Foundry', 162–3.
29 Cookson, *Age of Machinery*, 96–7, 101. Smiles describes a 'little colony of Scotsmen, mostly from the same neighbourhood' at Chowbent: *Industrial Biography*, 321.
30 Cookson, *Age of Machinery*, 91–5, 262.

castings to the Middleton Colliery and its railway, from at least the 1760s, and it was presumably through employment at Masbrough or another of their sites that Gothard learned of the excellent prospects for a small cupola furnace in Hunslet. He may even have been enticed there by colliery managers who knew him as an agent of the Walker partnership. The arrangement was all very neat, a foundry site with its own siding off the waggon-way.[31] Jubb was a later arrival to Leeds, living on the outskirts by 1778, possibly encouraged by that Wortley link to Kirkstall Forge and other sites, or perhaps by Gothard, whom he knew.

The south Yorkshire pioneers, then, were encouraged towards west Yorkshire by work contacts and family networks. They brought, or soon called upon, their own associates to join the new ventures. Workforces were built by attracting wider family or other connections from south Yorkshire, and training up promising locals.[32]

Migration did not mean an end to closeness. Indeed, it can strengthen and renew links between districts and individuals, as evidenced here. Joshua Wordsworth, a carpenter recruited for Leeds in 1803 by the Drabble brothers (themselves hired in by their relative Jubb) after Wordsworth had married their cousin, would later bring his own nephews from the home district. He also supported his two brothers, disabled and poverty-stricken former wire-drawers still in the Tankersley parish.[33] Richard Hattersley maintained strong business links thirty years after leaving south Yorkshire, buying specialist goods and steel from suppliers in and around Sheffield.[34] One such contact intervened during the crisis which followed Hattersley's death in 1829, rescuing the sons from insolvency and enabling the business to survive, and thrive.[35] These relationships were strong, extended beyond trade into personal life, and carried into the following generations.

Partners and ventures

The importance of the 'scene' – the workshop activity which fed innovation and influenced the physical shape of new industries – will be considered later. What, though, of the more formal institutions which sat above those shop floors, the

---

31 WYAS Leeds, WYL899/222, Middleton Colliery, tradesmen's bills; Cookson, 'Hunslet Foundry', 158–60, 153–5. There is record of Gothard buying iron at Wortley in 1740: BALS, SpSt 60513, Cope to Spencer, 10 Jan. 1739/40.
32 Cookson, *Age of Machinery*, 270–1.
33 Hey (ed.), *Militia Men of the Barnsley District*; Cookson, 'Joshua Wordsworth, 1780–1846', DNB.
34 WYAS Bradford, 32D83/2/3, bought ledger 1818–26.
35 WYAS Bradford, 32D83/33/1-2, letters to George Hattersley, 1832–3. See Cookson, *Age of Machinery*, 249–50, 117–21.

companies through which ironworking was organized and managed over the course of the eighteenth century? How those evolved shows an advance to meet changing circumstances, from Crowley onwards.

Before and after 1750, ownership and management models in iron production were remarkable. Mineral rights, charcoal forests, and ironworks sites were usually owned by great landed estates and managed by agents. Proprietors generally, though not always, remained at arm's length. Successful leaseholders could themselves accumulate great wealth. Questionable business practices and ruthless conduct were not unusual, some of that motivated by acute shortages of charcoal. The era witnessed networking at its most grasping and malign, energies spent defending narrow interests to exert control, build cartels, and shore up practices favourable to cliques.

Evans, Jackson and Rydén suggest that the British iron trade was in 'organizational flux', a 'restless fluidity', for more than a century after the Restoration, a situation forced upon it by steep rises in bar-iron imports from Sweden. The large ironmaking partnerships – most notably in the west Midlands, associated with Foley and Knight, and the Yorkshire-based Spencer groupings – gave a 'strong dynastic flavour' and a 'rather misleading solidity'. They were not firms as we know them.

> The various stages of the productive sequence (mining, smelting, refining, processing, manufacturing) were not internalized, much less integrated, within a single organization. Furnaces, forges, and processing mills were often only loosely articulated with one another.

The groupings were 'loose, ad hoc commercial alliances' within which goods and semi-finished iron products moved around. Networks of production 'shifted in kaleidoscopic fashion'.[36]

The Spencer combines

The Spencer and Foley arrangements, which from the start spanned regions, descended from opportunities opened up during the Commonwealth era. With this came a revival in ironmaking, and the construction of new works. The first John Spencer (c. 1599–1658) arrived from Shropshire to manage his Yorkshire holdings, accompanied by associates including Thomas Dickin and a cousin, William Cotton, as agents and works managers. The Spencer combines, based in and around Silkstone and Tankersley, peaked around 1730, when the family and their partners oversaw combines embracing at least 10 furnaces, 18 forges, 5 slitting mills, as well as ironstone mines and charcoal leases, in south and west

---

[36] Evans, Jackson and Rydén, 'Baltic Iron', esp. 643; for organization of Swedish ironmaking, also Evans and Rydén, *Baltic Iron*. See ch. 2, above.

Yorkshire and Derbyshire. They were also invested in Cheshire, Tyneside, and elsewhere, some of this in partnership with Denis Hayford.[37]

A façade of country-house respectability had been built on sharp business practices which at times strayed into criminality. A series of ironworks acquired during and after the Commonwealth came at some cost to their former royalist owners. Heirs had been swindled in order to take over other leases and premises. All in all, this represented a substantial transfer of ironworking assets in Yorkshire, by about 1680, into the hands of Hayford, William Simpson, Spencer, and their partners.[38]

Nine such syndicates, or combines, operated during the late seventeenth and early eighteenth centuries. They were bundled into groups, to include outliers in Yorkshire, north Derbyshire and east Lancashire, though the bulk was concentrated in south and west Yorkshire. The interests in Cheshire ironworking and the Derwent valley ran alongside these. Partnership structures were extremely complex, and the combines fluid, never static, with frequent trading in fractions of the holdings. The Rockley group, for instance, formed a syndicate of six ironworking sites. Ownership of the group was split 60:40, each section having several partners. These included sons and grandsons of the founders, and also of the ironworks clerks and agents with family and professional links to the original projectors, the Cottons and the Dickins. Other clerks and agents rose to junior partnerships and positions of responsibility in south Yorkshire: Spencer had the John Fells, father and son, and at Wortley was Matthew Wilson.[39]

Collectively, the Spencer combines were highly influential in how ironworking developed in Yorkshire and north Derbyshire before 1750. The groupings operated an internal market which embraced all stages of production, from mining ore to selling nails.[40] In a highly capitalized industry, this spread risk, phasing repairs and upgrades to furnaces and forges to protect overall capacity. But Spencer iron, largely produced to suit the combines' own needs, became less attractive outside its sheltered marketplace.[41] The trade's more general problem was the heavy reliance on charcoal, its scarcity evident as early as 1700. Not only was demand for iron rising, but there was a trend to turn wooded parkland to agriculture. Replacing mature coppice woodland took perhaps fifteen years. Spencer associates joined a cartel, with others, to regulate bidding for woodland leases, so beating down charcoal prices and sharing available

---

37   Above, pp. 21–2. For background on the Spencers and their associates, see Awty, 'Spencer Family, *per. c.* 1647–1765', DNB.
38   Awty, 'Denis Hayford', DNB; 'Charcoal Ironmasters', 83–4; Hopkinson, 'Development of lead mining', 199–200; Cookson, 'Wortley Forge'.
39   Raistrick and Allen, 'South Yorkshire Ironmasters', 168–9, 171, and *passim*; Cookson, 'Wortley Forge'. For the Fells, see Hey, *Fiery Blades*, 172–3.
40   Hopkinson, 'Development of lead mining', 199.
41   Raistrick and Allen, 'South Yorkshire Ironmasters', 172–7.

stocks. This agreement, founded in a milieu of suspicion, was on occasion breached. The relentless negotiating must have sapped energy and focus. But the cartel persisted, and by 1727 almost all the charcoal-iron industry within their region was concentrated 'in the hands of a very small group of men'.[42]

The uneasy allies worked together to squeeze out competitors. When Samuel Walker tried to lease Wentworth woods and ironstone measures near Masbrough in 1760–1, the cartel lifted its own purchase-price limit in order to disrupt Walker's progress in the charcoal-iron trade.[43] The Spencer combines themselves showed little taste for updating their own plant or taking on risk, even as early as 1720 when cornering the regional market in charcoal. Losing ground technologically in their own operations perhaps explains the fall in external sales of Spencer syndicate iron.[44]

The Spencer group – or more accurately their agent, the younger John Fell – did make efforts to track emerging technologies in the 1750s. Fell sent his blast furnace manager from Chapeltown to Coalbrookdale in 1759 'to learn how to blow with ground coals', but trials afterwards in Sheffield were unsuccessful. Neither did similar attempts to smelt with coke at other Spencer sites in south Yorkshire and Derbyshire, 1755–60, fare well, other than at Staveley, near Chesterfield.[45]

By then, any initiative from the Spencer family itself had ceased. The degree to which the owners relied on their agent (who was also a partner in the business) is clear from what happened after Fell died in 1765. William Spencer's one surviving son, John IV (c. 1718–75), had inherited in 1756. Educated at Winchester and Oxford, he had neither the training nor the appetite for business. With surviving partners unable to agree a way forward following the agent's death, John Spencer IV announced that he did not 'propose being any longer interested'.[46]

But already, men from the same group as Fell – experienced individuals who had managed the trade, agents and clerks who were sometimes also junior partners – were taking control of many of the former syndicate sites, negotiating directly with landowners for leases in their own names.

---

42 Hopkinson, 'Development of lead mining', 190–1, 195, 200, 208–11; and for syndicates in lead mining, in which many Spencer associates were also invested, 159–60. See Raistrick and Allen, 'South Yorkshire Ironmasters', 173, for 'coppice wood', or cordwood; and ch. 2, above, for charcoal.
43 Awty, 'Charcoal Ironmasters', 100; Hopkinson, 'Development of lead mining', 212; Raistrick and Allen, 'South Yorkshire Ironmasters', 177.
44 Raistrick and Allen, 'South Yorkshire Ironmasters', 177.
45 Hopkinson, 'Development of lead mining', 197–8.
46 Hopkinson, 'Development of lead mining', 203–4; Raistrick and Allen, 'South Yorkshire Ironmasters', 177.

Mid-century shifts

The ironmasters coming to prominence in south Yorkshire from the second quarter of the century were not, then, newcomers. Their ventures were solidly grounded, the owners knowledgeable and proficient in day-to-day management, technologies, and the wider trade. With autonomy came more flexibility than had previously been possible for them. Lack of finance may have been a restraint, at least at the start. But their own networks, connections to support and sustain them in many ways, proved to be important assets.

Matthew Wilson (*c.* 1676–1739) had been brought into the Wortley Forge cluster as a trainee in the 1690s by his brother-in-law, John Spencer III. He worked his way to managing partner at Wortley and Dodworth. At Wortley, Wilson rebuilt both forges in 1713, and commissioned a new wire mill, 1727–8. He was a partner in the Thurgoland wire and slitting mills, with shares in several forges and mills elsewhere in Yorkshire, all alongside Spencer family members and other brothers-in-law.[47] Wilson bridged the transition of that older model into something new. In selecting and training a nephew, John Cockshutt (1711–74), as his own heir and successor, Wilson secured the future of Wortley Forge into a new era.

Aside from any technical or commercial matters, the Spencers experienced two failings common in family firms. One was embodied in John Spencer IV, the last in the line, who did not care whether the company survived. The other was the bad blood between family members and associates, and then, the Spencers having discarded them and filled supporting roles with in-laws, allowing those relationships too to sour. There was demonstrable lack of trust, even among relatives who knew each other well and whose predecessors had collaborated over many years. The Spencer story serves as a counterpoint to networks perceived, rightly or not, as high-trust – the Darby ironmaster circle, certain natural philosophers, and workshop-based engineering, for instance.

As the Spencers themselves faded as a force, Matthew Wilson became increasingly indispensable, and aware of his own worth. The Spencer crisis came when John III died in 1729, soon followed by his younger son Edward. Edward, at odds with his brother William, bequeathed his half-share in John Spencer III's ironmaking interests to Matthew Wilson, who was their uncle as well as works manager. The disputed property was worth a fortune, estimated in 1738 at £32,197. William, already very wealthy, was infuriated.[48]

The truculent Spencer refused to acknowledge his brother's will, for a decade pursued a case against Wilson, and, after Wilson's death in 1739, acted aggressively towards Cockshutt. Various claims and ruses were used, including reducing the value of contested properties by loading them with debts

---

47 The relationships are described in Cookson, 'Wortley Forge'.
48 Cookson, 'Wortley Forge'; Hopkinson, 'Development of lead mining', 204–5.

from other syndicates. Nor would Spencer cede the lease of Wortley Forge to Cockshutt, and he blocked access to the forges' usual sources of pig iron.

Remarkably, through this long period of strife, William Spencer and Matthew Wilson (then, from 1739, John Cockshutt) remained partners, along with other close relatives, in this portfolio of ironworks. Wilson had age and authority on his side, but, anticipating that greater trouble lay in store for his young heir, provided in his will for two 'supervisors' to advise and help.[49] Cockshutt's letters to Spencer through the period of Wilson's illness, and after Wilson died, display an increasing self-assurance and a distaste for Spencer's methods. One partnership had liabilities from, as Cockshutt called it, 'the nail affair'. He agreed with Wilson's view, that creditors would be satisfied by a handover of remaining stock, which had been undervalued by Spencer. A widowed aunt had been tricked into handing over her share of stock, without being discharged from her share of liability for any other claim against the company. Cockshutt sought a guarantee that no other partner would be thus deceived:

> every penny I have ... from company's stock I shall apply to the company's debts – but if I can't be trusted to do that I don't see any reason why I should intrust anyone else without security much less without promise.

A bond was agreed, the stock's value to be assessed by arbitrators, and that matter should – though may not – have been settled in 1740.[50]

Here and afterwards, Cockshutt's lack of ready money was exposed, and so were the avarice and pettiness of his wealthy cousin Spencer. For years their bitter dispute continued, over control of Wortley Forge and other sites. Yet all the while they were obliged to connect over everyday matters of business. How was this possible? As Wilson's assistant, Cockshutt had been prepared to 'wait on' Mr Spencer at Cannon Hall. He still called on Spencer, and wrote him letters, but from 1739 increasingly on his own terms. Messages passed via Thomas Cope, Spencer's agent at Wortley Forge, and other intermediaries. 'Mr Cockshutt will be with you on Saturday & not before'.[51] Spencer routinely withheld money owed to Cockshutt, who had reopened wireworks and slitting mills at Wortley and supplied the forge. In response, Cockshutt controlled access to workshop records and accounts. Yes, he told Spencer, you and other partners may examine the books, but we will retain them 'untill things be settled, being (till then) obliged every day to have recorse to them'.[52]

---

49 TNA, PROB 11/699/19, will of Mat[t]hew Wilson; BALS, SpSt 60512, official copy of Wilson's will (26 Feb. 1738/9).
50 BALS, SpSt 60512, Cockshutt to Spencer, 16 June 1739, 28 July 1739; Bond 8 Nov. 1739. See also Hopkinson, 'Development of lead mining', 203–5.
51 BALS, SpSt 60513, Cope to Spencer, 10 Jan. 1739/40.
52 BALS, SpSt 60512, Cockshutt to Spencer, 26 May 1740, from Wortley wireworks.

He complained in 1740 that Spencer caused him 'needless work' by sending untrained clerks to inspect the books, and always reminded him of the unpaid slitting bills.[53] Later that year, Cope informed Spencer that 'Mr Cockshutt & Mr Thomas were at the slitting mill yesterday'. Having instructed the slitter William Allsop to stop working, they 'lockt the mill door & took the key away'. Cockshutt told Cope that slitting would resume once Spencer paid £80 that was owed. Cope reported back to his master that Cockshutt made light of Spencer demanding the key, 'says you have nothing to do with the Mill without his consent'.[54]

Even before the main dispute was resolved in 1743 by arbitration – Wortley Forge and four furnaces nearby awarded to Cockshutt, the Kirkstall and Colne Bridge forges to Spencer – Cockshutt was already busy updating and adding to parts of the Wortley cluster. After the settlement, he took the lease of Bank Furnace to secure supplies of pig iron for Wortley Forge, helped by a £10,000 loan from the forge's landlord Edward Wortley, and a new partnership with the Sheffield nail merchant and banker Joseph Broadbent.[55] Thomas Cope, it appears, continued for a short time at Wortley Forge. Having been seen as favouring Cockshutt – and perhaps made scapegoat for his former master's loss of the Wortley properties – he was pursued with legal threats by Spencer and his associates.[56]

The Spencer–Cockshutt story is one of an unequal family partnership turning from sour to hostile. It may be an extreme case. Yet – while the balance of power and assets was loaded in Spencer's favour – Cockshutt was not intimidated, asserted his rights, used the law to protect his position, and was ready to endure some uncomfortable years in order to secure his inheritance. Matthew Wilson had prepared him for confrontation, and introduced him to many of the individuals who helped protect the younger man, in some cases by their very impartiality in the face of the mighty Spencers. For it was in the interests, generally speaking, of these various agents, intermediaries, and merchants, and others who formed the infrastructure of the region's iron trade, to uphold its integrity. John Cockshutt's clash with Spencer was not a failure of community. That had been personal, a loss of trust, the need to deal with an unreasonable man. But it was the coming together of voices of reason and equity which rescued the situation.

Though still powerful, the Spencer groupings were losing their regional dominance. The 'intermediates' – men of broad social equivalence to the

53 BALS, SpSt 60512, Cockshutt to Spencer, 2 June 1740.
54 BALS, SpSt 60513, Cope to Spencer, 18 Aug. 1740; 21 Aug. 1740.
55 Hopkinson, 'Development of lead mining', 204; Cookson, 'Wortley Forge'.
56 BALS, SpSt 60513, Cope to Spencer, /81, 19 Feb. 1744/5; Spencer to Cope, 16 Oct. 1745; Jonathan West, Cawthorne, to Cope, 13 Mar. 1745/6; /127 Cope in Nantwich to Benj. Dutton, Cannon Hall, 26 Dec. 1747.

Spencers, though without their wealth – came to the fore. Responsibilities were devolving to a professional class of managers, technical advisers, and record-keepers.[57] Why did this occur? Older business formations had run their natural course, it seems. Technological matters were pushing change, as demand for iron continued to rise. The mid-century, lacking the clarity of any retrospective view from 1800 or later, was searching for its own pathways. Pressures to improve ironworking methods brought turbulence, with only limited success.[58] In Yorkshire, at least, a weakening old-style leadership had to concede to new approaches.

'Intermediates' sums up the network which acted to John Cockshutt's advantage. Together, this amounted to a protective framework of law, protocol, experience, and knowledge. Cockshutt had suggested Spencer should better train his clerks, the ironworks administrators, complaining of one sent to inspect the wireworks books that

> If you take clerks as apprentices & want to give everybody this trouble for their instructions I think tis more reasonable you qualifie them yourself or employ somebody that can.[59]

Cockshutt evidently expected higher standards. His close adviser, formalized by Matthew Wilson's will, was Callisthenes Thomas, gentleman and specialist in the iron trade, resident at Huthwaite in the late 1730s and afterwards of Wortley Forge. He had been Cockshutt's accomplice on the operation to assert control of the slitting mill in 1740. Probably Thomas – like John Tyas of Monmouth who applied to be clerk to Spencer in 1740, and several other workers at the forge in this period – originated in south Wales.[60]

Thomas was clearly aligned with Cockshutt's interests – later, he married Cockshutt's sister – but over the course of his career in and around Wortley he also represented other ironmasters and interests, including the Cotton family. On behalf of a partnership including Spencer, Cockshutt, John Watts, and John Fell, Thomas was appointed to manage timber, wood, and bark sales after the death of James Oates (an uncle of both Spencer and Cockshutt since marrying Catherine Wilson, and partner in several ventures). Messages exchanged on that same matter reveal Thomas as a master of style and script in comparison

---

57   See 'intermediates', in ch. 4, esp. p. 98.
58   For reasons discussed in ch. 2, above.
59   BALS, SpSt 60512, Cockshutt to Spencer, 2 June 1740.
60   BALS, SpSt 60516, including Spencer to Oates, 10 Jan. 1738/9; Cookson, 'Wortley Forge', 64.

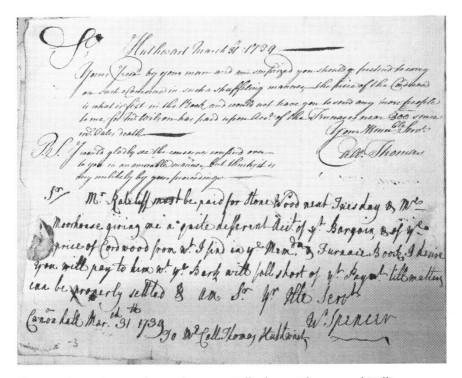

Plate 11. An exchange of notes between Callisthenes Thomas and William Spencer, 1739, in the aftermath of another Spencer partner's death. Thomas, an ironworks agent with the rank of gentleman, was evidently undaunted by Spencer's great wealth and power. Spencer's reply is surprisingly scrappy. Source: BALS, BA SpSt 60516/3.

with William Spencer. He was also ready to assert himself, calling out Spencer's manipulation and obstinacy.[61]

In this, Thomas was not unique in his profession. Other agents managed to be closely associated with one ironmaster and yet remain frank and objective. John Fell of Attercliffe, Spencer's representative, was named as arbitrator in the bond which was to settle 'the nail affair', 1739–40. Alongside John Watts of Kirkstall, he would value a stock of unsold pig metal, bar-iron and rod iron at Wortley Forge, assessing its worth when Spencer entered that partnership. Thomas witnessed the bond.[62] Similarly, the nail merchant Joseph Broadbent was everywhere, greasing the wheels of trade, striking bargains, advising, investing.

61 BALS, SpSt 60516, 1 Apr. 1740; 60516/3, Thomas to Spencer, with reply, 31 Mar. 1739 [presumed 1740 NS]; Cookson, 'Wortley Forge', 62–4.
62 BALS, SpSt 60512, Bond 8 Nov. 1739.

Being aligned was only a part of these men's portfolios. Expertise on that level – bear in mind their increasing level of technical understanding, and the gathering speed at which the trade evolved – was a valuable asset, to be cultivated and respected. A reputation for poor judgement could render an agent unemployable, even within a family business, which is what happened a generation later to Cockshutt's third son, Edward.[63] Protecting the integrity of iron-trade skills, workshop-based or intellectual, was in everyone's interest.

John Cockshutt's initiative and fresh approach contrasted starkly with the Spencers' risk-averse fading businesses. He and his sons reconfigured plant and sites to suit new products and technologies, funded by hefty borrowing and partnerships. They accepted financial stringencies, meaning that the sons could not always be employed at Wortley. Those absences became a positive, broadening minds and introducing valuable contacts from the world of science and engineering.[64] But the Cockshutts were far from any personal hardship, being comfortably situated, educated and trained for the task, technically able, outward-looking, confident, and curious. The three Cockshutts were able technologists, innovators on a modest scale, and thoroughly understood their business. By the time that William Spencer's disengaged heir, John Spencer IV, announced his withdrawal after John Fell II died in 1765, the Cockshutts were well-placed enough to assemble their own syndicate to take over the Duke of Norfolk's Ironworks. Other small ironmasters followed suit with former Spencer group sites across the wider region.[65]

In a neighbouring parish and at about the same time that John Cockshutt took control at Wortley, Samuel Walker and his brothers, of Grenoside, Ecclesfield, began trials to remelt iron in furnaces they had constructed themselves. Earlier generations of Walkers had been farmer/nail-makers. While the family was quite comfortably situated, far better than many in the trade, the brothers saw little future in their father's occupation. When Joseph Walker died in 1729, the eldest son, Jonathan, still only eighteen, took on managing the farms. After the death of their mother, Anne, Joseph's second wife, in 1741, several properties passed to the three individually. Within weeks, the experiments started – thought to have involved brass as well as iron, and using a reverberatory (air) furnace from the beginning, or almost so. At first it was mainly the youngest, Aaron, and their half-sister's son, John Crawshaw, trying to cast in pots, 'but with bad success'. Meanwhile Jonathan farmed, Samuel taught in the village school, worked as a surveyor, and made sundials, and Aaron did odd jobs on

---

63   Cookson, 'Wortley Forge', 69–70.
64   Cookson, 'Wortley Forge', 67–8.
65   Above, pp. 59–60.

farms and sometimes made nails. Inheriting their mother's rental income had made the furnace trials possible.[66]

The project found success once Samuel joined Aaron to modify the furnace, with advice from a former foundry worker, Samuel Saint, about producing small castings. By the end of 1741 they were in production, selling five tons of castings the following year. Aaron could take a full-time wage, Crawshaw was employed as needed, and Samuel worked at the furnace when not teaching. In 1744–5, other employees were taken on when required, and Samuel quit his job. Jonathan joined his brothers as partner in 1746, his talent as transport manager finding a place. The Walkers showed that it was still possible, in this trade that demanded relatively high levels of capital, for a start-up business to enter and grow on reinvested profits.[67]

Samuel Walker was the driving force, ambitious and informed, but the brothers' combined aptitudes gave them an edge. Aaron, seeking a way out of the moribund local nailing trade, sparked the venture, Samuel's insights made it viable, and Jonathan managed logistics. Their readiness to take a strategic view moved them to the next level: while the roads around Grenoside were relatively good, and they continued on that site, the Don Navigation's arrival into Rotherham convinced them to focus on Masbrough in 1746, just as the business took off. John Crawshaw was then a partner and investor, and John Booth, nail chapman, joined as part-financier of a steel furnace in 1748. Subsequently the company broadened its activities into a range of cast- and wrought-iron merchandise. After 1770 they became known as suppliers of cannon to the Royal Navy, and of cast-iron bridges, and made a lucrative move into the lead industry.[68]

When the last of the founding brothers, Samuel, died in 1782, the firm was valued at £100,000. Aaron Walker (d. 1777) had managed casting and steel-making within the works, exhibiting 'more ingenuity than patience'. Jonathan, who died the following year, was noted for caution and prudence in managing 'our farms, teams, roads, &c'. The 'irreparable loss' of Samuel,

---

66 Morley, 'Walkers of Masbrough', 1–6 and *passim*; John (ed.), *Minutes relating to Messrs. Samuel Walker & Co.*, i, and *passim*; RA, 1263-Z Munford typescript; Walker family ephemera, vols 1 and 2, including transcripts of inventory and will of Joseph Walker, d. 21 Dec. 1729; annotated pedigree of Walkers; BIA, Probate of Anne Walker of Ecclesfield, July 1741 vol. 87, f. 433. Foster misreported the mother's death as 1740, but her will and probate, and the Ecclesfield burial record cited by Morley, confirm it as 28 Feb. 1741.
67 Morley, 'Walkers of Masbrough', 4–9; RA, 1263-Z Munford typescript; John (ed.), *Minutes relating to Messrs. Samuel Walker & Co.*, 1–4.
68 Morley, 'Walkers of Masbrough', 2, 6–9; RA, 1263-Z Munford typescript; John (ed.), *Minutes relating to Messrs. Samuel Walker & Co.*, i–vii; Chrimes, 'Walker Bros', BDCE. Samuel jun. married the widow of John Booth's son, and made a small bequest to the daughters of that earlier marriage: RA, 31-F/3/1, copy probate, 1792. Booths remained close to the Walkers in various joint enterprises for the duration of the company: John (ed.), *Minutes relating to Messrs. Samuel Walker & Co., passim*.

with the 'integrity, industry, foresight, and perseverance, which appeared in all his actions', was a bitter blow.[69]

Considering the dangers facing a valuable family business on losing its dominant character, the Walkers managed well. The next generation was already within the partnership: Samuel's sons (Samuel jun., Joshua, Joseph, and Thomas); and Jonathan jun., who was married to his cousin, Samuel's daughter, and was considered a brother. 'John Crawshay, or rather Crawshaw' remained in the firm after four decades. Fragmentation was largely avoided, but the partners acted decisively when necessary. In 1782 Samuel jun. bought out Aaron's only son, John (1760–1804), a partner for at most two years whose 'imprudent course of life had brought him into ... many difficulties'. That share was divided among the four brothers. Then, in 1789, as Crawshaw was about to hand over his seat on the board to his son, the Walker members, citing their disgust at 'the cold unpromising disposition and conduct of Mr John Crawshaw junr', announced their intention to end that partnership. The Crawshaws, both still active in works management, then withdrew.[70]

How exactly the 'extensive and multifarious' Walker concern had operated under the three brothers is not clear. Most likely the younger generation trained in specific areas of management before progressing to partnership. Specialist agents, effectively subcontractors, managed the workforce. A year after Samuel's death, an unusual and significant document, now believed to be the work of his youngest son, Thomas, drew a management plan for the business in its new shape. 'It is not possible', he began, 'for every partner to attend in so minute a manner to every branch, as a master ought to do.' He discussed the use of agents, cautioning against over-reliance on them and the necessity to be diligent against their 'mischiefs'. Of these, a long list followed. At the booklet's end – along with practical suggestions about minor matters – Thomas laid out democratic principles: that each partner take a lead in certain departments while also retaining an overview, that everything was up to discussion and partners speak freely. 'Every Wednesday evening shall be continued the time of our meeting together as usual'. That confirms existing practice, and other sections of this document probably describe routines already in place.[71]

Thomas wanted to convince his co-partners to follow the plan and his approach. The balance between overview and over-interference must be found, he suggests, through a spirit of friendly communication:

69 John (ed.), *Minutes relating to Messrs. Samuel Walker & Co.*, 13–14, 18–19.
70 Morley, 'Walkers of Masbrough', 30; John (ed.), *Minutes relating to Messrs. Samuel Walker & Co.*, 20, 22–3. John Walker is said to have died in Russia: Hey, 'Walker family', DNB.
71 RA, 328-B/1/1. The booklet is catalogued as 'Proposals by Joshua Walker, c. 1785'; Munford (RA, 1263-Z) shows why this date cannot be correct, and identifies the author with near certainty as Thomas Walker, late 1783 to early 1784.

No man in any line of life can ever do good if he has not an emulative mind, and has room given for this very proper passion to expand. This cannot be the case as our business is now managed, as we are apt to dip indiscriminately into the different branches of our concern, of course the issue of any plan or scheme can scarce be charged to the account either of prudence or neglect.

And then, reflecting the family's Methodism, comes something which would be incredible in modern business planning:

Surely none of my partners think that the principal object in this world is *money* – that our dividends, through the assistance of providence, prove handsome, that we need not mind whether our behaviour to each other is manly, unreserved and forgiving, or the contrary – or that, it is of no importance that our pursuits are creditable in the eyes of those whom we esteem good, and sensible. I know if I ask each of you apart what your wishes are, I should have a reply which would do credit to your heart & understand[in]g, and upon this certainty I found my hopes that our business will be conducted upon the most happy and beneficial plan.

The management structure he proposed, following in an appendix, defines responsibilities for each of the six partners, with a named agent for every department. For Joseph, as an example, the remit was woods, colliery, ironstone pits, pottery premises, white lead works and accounts, and building alterations, with four agents in support. Were Thomas's recommendations followed? It is likely, though perhaps with amendment. Adjustments must have followed the Crawshaw withdrawal – they had been concerned in furnace and foundry departments – and further changes of personnel and plant.[72]

In many ways, the Walker company was extraordinary; in their own way, so were the Cockshutts. Those enterprises joined a trade which stood on a threshold, while the Spencer interests receded. They made steady progress and showed ways forward for their industry. Others, too, were engaged in the reinvention. From the 1740s, the new wave of ironmasters explored novel, outward-looking, hands-on ways to do business and make iron. Kirkstall and other sites allocated to Spencer in the 1745 agreement found new owners and were in due course regenerated. John Wilkinson took on the lease of Bersham Foundry, near Wrexham, in 1763, and converted it to cannon- and cylinder-boring, famously supplying Boulton & Watt. Other ventures were altogether new: the Walkers, Charles Gascoigne and his Carron associates, Hunslet Foundry's small cupola furnace serving specific local demands.[73] South Bradford opened up for Dawson and his Low Moor partners, and to other new ironworks, after 1780 with the arrival of waterways. The Darby company,

---

72 RA, 328-B/1/1; Hey, 'Walker family', DNB. Some family members adopted a Calvinist strand of Methodism.
73 Hayman, *Ironmaking*, 86–7.

it appears, remained in many regards an exception in overcoming recurring threats and troubles, surviving and staying relevant for almost a century.[74]

Partnership
Are any general conclusions to be absorbed from the above cases? All operated within aspects of the iron trade, and met with – at least partial – success. Several partnership models are evident, while other ventures apparently thrived with a single director, or a hybrid, one dominant individual among lesser partners. But over-dependence on one person, particularly in view of the life expectancy of eighteenth-century ironmasters, may not have been the wisest choice.[75]

Three times the Darby company lost its principal partner, its guiding light – four, if the *de facto* heir William Reynolds counts. Each resulted in severe difficulties. Although, somehow, the business endured, for periods it was becalmed and operating below par.[76] More than once it was rescued by Quaker associates, not because they were Quakers, but because they were family, friends, and fellow partners in a viable concern. Quakers have been caricatured as benign, serious and businesslike, as though they shared one personality.[77] Certainly the Darby firm had its own way of working, and the partners formed a tight group sharing a culture and family connections. That could, though, be said of many ventures. Gerbner has demonstrated that Quakers at this time did not subscribe to a common set of ethical codes, even on slavery. Principles varied between continents and different Quaker meetings, and shifted over time.[78] Some English Quakers advocated an end to slavery before 1700, while enslaved people were still exploited by Quaker settlers in North America a century later. Many of the first Abraham Darby's patent iron pots, the foundation of his family's business, were exported to North America through Thomas Goldney, his close associate in Bristol. The Goldneys were complicit in the slave trade, and it is more than likely that they traded Darby pots to plantations abusing enslaved people. A product identical to Darby's original designs, called a 'Negro Pot', was still offered for sale by the Coalbrookdale company more than a century later.[79]

---

74 See ch. 2, above.
75 This is based on impressions and limited data. It may be that their equivalents in other industries were equally short-lived. Many ironmasters lived in unhealthy proximity to their works, the Darbys in the confines of Coalbrookdale.
76 See ch. 2, above.
77 A cliché which Ashton, *Iron and Steel*, 213–20, repeats.
78 Gerbner, 'Slavery in the Quaker World', shows that William Penn himself traded enslaved Africans from Barbados when founding Pennsylvania in 1682.
79 For Thomas Goldney and Goldney Hall see TNA: https://webarchive.nationalarchives.gov.uk/ukgwa/20210803123212/https://livelb.nationalarchives.gov.uk/pathways/blackhistory/copyright.htm. In 1727 the British Society of Friends Yearly Meeting banned owning or dealing in slaves, but in North America only in 1774 were Quakers with involvement in slavery finally instructed to stop or to leave the Society: https://www.quakersintheworld.org/quakers-in-action/11/Anti-Slavery.

The point here is that Quaker companies did not match a template later laid upon them. In understanding why companies behaved as they did, the wider business environment could be more influential than the owners' Quakerism.

Quaker merchant banks came to the Darbys' rescue more than once, and underage heirs were helped by astute relatives. But this – support from within commercial cliques, especially ones with investments in common – was not a feature only of Quakers. Neither were such facilities extended to all Quakers. Arguably the Coalbrookdale owners found greater affinity with non-Quakers of their own social status than with Quakers of lower standing. Extending business networks through boarding school acquaintances, and marrying outside their home region, was common also for Unitarians, and Anglicans for that matter. Nor was a shared faith a guarantee of trustworthiness, or efficiency, in any of those instances.[80]

The final flourish at Coalbrookdale was spectacular, an ambitious new approach launched by William Reynolds in the 1790s. Seeing the transformation that puddling brought to the iron business, and anticipating Shropshire's eclipse by south Wales, he parted ways with his Darby relatives. Able and ambitious, with serious technical and chemical interests, Reynolds had a bigger vision. Coalbrookdale's difficult topography – restricted sites, water shortages – impeded further development. Reynolds sketched, and started to build, a new model of industrial integration downstream of the dale, where the Severn headed south, including a canal centred on Coalport. The design extended beyond iron and coal, into china, tiles, glass, brick, tar, and sulphuric acid, connected by an elaborate transport system. By-products of one process would become constituents of another. Reynolds' schemes started to drive a revival. A small river terminal connected the Shropshire Canal (and beyond) via an inclined plane to the Severn, in 1791. This transported Shropshire coal and iron, and attracted new industry (in which he heavily invested) near the waterway. Had Reynolds lived longer, he might have achieved something truly great. But he died at forty-five, in 1803, with the concept not fully consolidated, unfinished.[81]

On such chance, the success of a venture hinged. Losing an owner could mean disaster, but was difficult to protect against. Taking partners had advantages but came with its own risks. That explains, at least in part, the tendency to build from the familiar and work with people who were known, while avoiding others known to be unreliable. That was true even for the most ground-breaking

---

80 Cookson, 'Quaker Networks', esp. 172–3.
81 Hayman, *Ironmaking*, 98–9; Milligan, *Dictionary of British Quakers*, 139–40, 357–9; Bruton, 'Shropshire Enlightenment', 60, 102–3, 106; Powell, 'William Reynolds (1758–1803)', BDCE; Baggs, Cox, et al., *History of the County of Shropshire. XI. Telford*, 21–3, 40–56; Porteous, *Canal Ports*, 46, 49. See above, ch. 2, for the Darbys; pp. 119–20 for Reynolds' scientific interests.

enterprises. It was pragmatic, a way to protect personal interests before the advent of limited liability companies in the 1850s.[82]

The closely linked matters of raising finance and delivering desired outcomes were reflected in partnership agreements. Dishonest, incompetent or fractious associates were not the only threats. Companies could be sunk if a partner became insolvent through failure of other business interests, as happened to the Arkwright-style Kirk Mill, Chipping, when the Lancaster merchant house which sponsored it failed in 1787.[83] Illness or untimely death were an unpredictable element in any business. Hunslet Foundry was almost finished at the start when the main financier, who was also an active co-partner, died in 1771, aged thirty, after less than two years in business.[84]

But technical expertise was increasingly recognized as a business asset. It qualified Timothy Gothard to enter the Hunslet Foundry partnership, though he could invest little cash. Many aspiring partners had technologies or skills to share, but no funds; those with cash, and perhaps an idea or a patent licence, needed workshop skills. Thus textile innovators partnered with technicians, like Lewis Paul and John Wyatt in the 1730s, Paul's concept and money combining with Wyatt's mechanical talents to produce a pioneering version of spinning by roller.[85] The deal struck between Boulton and Watt in 1775 took account of Watt's patent rights and Boulton's far superior material assets.[86] William Carr was made partner in Kirk Mill in 1785 because he knew the Arkwright technology. Not being a cash investor, he avoided financial ruin when the main investors failed, and briefly continued to run the factory until it was sold.[87] Charles Bage received a minor share in Ditherington Mill as remuneration for having designed it in 1796.[88] There were also many inactive partners in these ventures. William Lister entered a partnership purely to part-fund a large expansion of Matthew Murray's works in 1804, afterwards proudly styling himself 'ironfounder'.[89] There was a point to this, beyond the cash: a respectable sleeping partner asserting his stake in a company also marked its substance. Other less active partners – and this appears a pattern in the iron trade – were merchants in associated activities, nail factors and iron dealers, part of the payback being to secure supplies of the business's merchandise. Those

---

82 Defoe's *Complete English Tradesman* is a litany of the disasters which threatened businesses.
83 Cookson, *Age of Machinery*, 89–90.
84 Cookson, 'Hunslet Foundry', esp. 155–8.
85 Cookson, 'Lewis Paul (d. 1759)', DNB. Thanks to John Styles for sharing his recent findings on the value of Paul's pinking-machine patent.
86 Tann, 'Matthew Boulton (1728–1809)', DNB.
87 Cookson, *Age of Machinery*, 89–90.
88 Swailes, 'Charles Woolley Bage (*c*. 1752–1822)', BDCE; Giles and Williams (ed.), *Ditherington Mill*, 1; and thanks to Joanna Layton for further information.
89 Cookson, *Age of Machinery*, 113, 278.

specialist merchants were also a handy source of advice to their fellow partners on commercial and technical matters.

Possibilities to progress in a business have been mentioned – clerks who were *de facto* managers, managers becoming agents or subcontractors. Collieries and ironworks were supervised in similar ways, usually on lease from landowners, with the workforce at a step removed, overseen by intermediaries. The Walkers adopted such a system of delegation, prompting Thomas Walker's later contemplation on the art of managing agents.[90] In smaller works, the chain of command was flattened, the owner close to the shop floor, the clerk often a relative, and perhaps servicing more than one business. Whether there were prospects of advancing to partnership, or becoming a partner elsewhere, would depend not only on how the company progressed, but whether the owner had suitable heirs to promote.[91]

James Cockshutt's experience illustrates the dilemma. As younger son, he stood aside from Wortley Forge, it seems without animosity, for as long as his brother was manager. James's technical interests aligned with the forge's business, but perhaps it was hard to accommodate him, or finances did not allow. He may simply have preferred to carve a career which included civil engineering. But after two decades in Wales, he was sacked after Richard Crawshay took over at Cyfarthfa. Being a partner there had not protected Cockshutt, or three others (including Crawshay's own brother-in-law), from dismissal.[92] A year later, in 1792, the civil engineer Watkin George, who had worked alongside Cockshutt in designing the Glamorgan Canal, joined Crawshay as partner. The skills formerly provided by Cockshutt were needed, and George went on to develop Cyfarthfa with bridges, aqueduct, two blast furnaces, and the 50ft (15m) diameter 'Aeolus' cast-iron waterwheel powering a blowing engine supplying blast to the furnaces.[93]

Owners' confidence determined how far innovation was supported. New ventures grew upon skills and knowledge which were specific, though often not fully defined or honed. Crawshay believed that he would do better by taking charge, and perhaps he was right. Collaboration was not in his lexicon. As partner, the clearly talented George, a man of humbler origins, may have been more tractable than the previous set.

---

90 See above, pp. 163–4.
91 For the duties of the ironworks clerk, see Hopkinson, 'Development of lead mining', 205–7; Hayman, 'Shropshire wrought-iron industry', 146; Evans, *Labyrinth of Flames*, 56–70.
92 Cookson, 'Wortley Forge', 67–9.
93 Chrimes, 'Watkin George, foundry manager', BDCE; Cossons (ed.), *Rees's Manufacturing Industry*, V, 360; Cookson, 'Wortley Forge', 68; Historic Landscape Characterisation, Merthyr Tydfil, 012 Cyfarthfa Iron Works: https://www.ggat.org.uk/cadw/historic_landscape/Merthyr_Tydfil/English/Merthyr_012.htm. See Plate 4, above, p. 54.

But ironworks management took many forms. The Low Moor Company's main investors made a seemingly odd choice in appointing a 50-year-old Presbyterian minister to run their new business on the fringes of Bradford. But Joseph Dawson had industrial experience, a knowledge of ironmaking and coal-working, and a good grounding in science. Supported by Presbyterian charities during his education at Daventry Academy and Glasgow University, in middle life a sizeable inheritance allowed Dawson to join the promoters of the Low Moor Company in 1788. His fellow directors, businessmen lacking technical knowledge, evidently allowed him freedom to plan the works, which were in blast from 1791. Dawson, well-read in chemistry, continued to observe and experiment with the performance of Low Moor's furnaces. From 1800, as first president of the Yorkshire and Derbyshire Iron-Masters' Association, he presented papers on iron-smelting and the composition of coal and iron. But he could not tempt those northern ironmasters to offer any contribution of their own. He was similarly frustrated in launching a Bradford Lit. & Phil., which fizzled out after two years through lack of support.[94]

For Dawson, then, it was not all about making a fortune, nor even a race for industrial innovation, although he did those things too. He had a broader remit, a devotion to science, to botany and the prized minerals and fossils emerging from Low Moor's depths, which he encouraged the miners to bring him. These were shared with botanists, geologists, and collectors, local and international.[95] There, it seems, is a point in common with Thomas Walker of Rotherham. For surely there was more to this world than money?[96]

Spencer, Cockshutt, Darby, indeed all ironmasters, were obliged to manage innovation, one way or another. They must find their own path, appropriate to the venture, and suiting their own appetite for risk and upheaval. The choice could be to take little initiative while making the most of a declining asset, as with end-phase Spencers. For the newly independent, though – Cockshutts, Walkers, Dawson – accommodating substantial innovations in methods, resources, markets, and more, was fundamental to the mission. They planned for the longer term, and were technologically equipped for that task, or at least better so than most. The Carron Company was different, the ultimate example

---

[94] The legacy was in 1775 from a Leeds clothier and fellow member of Mill Hill Chapel, Christopher Holdsworth, unrelated to Dawson: BIA, Proved at York, Prerog. 1775. Holdsworth probably influenced Dawson's appointment to Idle Chapel. See also *Fortunes Made in Business*, 89–128; Pacey, 'Emerging from the Museum'; Morrell, 'Wissenschaft in Worstedopolis', 8. Research into Dawson's career and circumstances surrounding the Low Moor Co.'s establishment continues. Thanks to Mary Twentyman for much additional information.

[95] Pacey, 'Emerging from the Museum'. Dawson's will notes scientific instruments, an extensive library, and his mineral collection (now housed in Cliffe Castle Museum, Keighley): TNA, PROB 11/1558/445.

[96] See above, p. 164.

of heavyweight networking grounded in scientific curiosity. Yet it was almost derailed, perhaps by hubris, and certainly through financial and managerial failings.

Any model for a successful, innovative, eighteenth-century business start-up would feature some of these qualities: resourcefulness, fairly obviously; a grasp of relevant technology and merchanting; an appreciation of the market and its potential for new products. Often there was a self-assured young man supplying the impetus, a figure bringing an idea, possessing a measure of self-sufficiency perhaps linked to a recent inheritance. Access to finance, borrowed or otherwise, was important, and so, to meet the inevitable setbacks, was creditworthiness. In all of this, the new undertakings must have connections, for support, information, and other needs. Some of the ventures – whether partnerships, associations, or individuals – caught a wave of innovating, and their creativity spread and endured.

Those generalizations are not confined to iron and textiles. Whatever the industry, a promising start could not guarantee a good outcome. Samuel Crompton had neither means nor knowledge to protect his brilliant concept, and quickly lost control of the spinning mule. The famed projects which culminated in great businesses tended to be better resourced, often involving merchants. In Leeds textiles, there were Benjamin Gott's organizational feat in woollen-cloth production at Bean Ing, an adjunct of his merchanting business; the Elams, applying their expert knowledge of trading with North America to cloth exports; John Marshall, in effect launching the mass production of linens in factories. All were merchants, but Marshall alone focused primarily on mechanization. He and Gott shared the profile of confident young man bequeathed a fortune and full control of a merchant house. The Walkers and Cockshutts, despite similarities of place and time, differed in background, wealth, and how their businesses were managed. The Walkers' more modest start culminated in a vast enterprise. There was no problem accommodating multiple offspring, and having young and imaginative partners was likely to bring vitality and regeneration.

The impact of luck, good or bad, is never to be underestimated. It was not only partners and family members who influenced businesses for better or worse. Members of the workforce, in varying relationships to owners, developed concepts or brought ideas into solid form, and thus made a difference. Matthew Murray did it for Marshall; David Mushet, for the British iron industry more generally when his own employers tired of his innovating. William Murdock could have done it for Boulton & Watt, had they allowed it. But the dozens of forgemen, largely unidentified, whose skills brought puddling into practical use in their own workshops are representative of those anonymous legions who ultimately made things work.

7

Creating a scene

A noble scene of industry and application is spread before you here.

Defoe, *Tour thro' the Whole Island*, Letter 8, Part 4

The circles of influence, those innovative formations at work in various settings – how are these to be captured and described? So much was ad hoc, and hard to pin down, even as the groups themselves are brought into clearer focus. The 'scene', for want of a better term, grew within different contexts. For industrial purposes, that could include technical knowledge, expertise and experience, science, organizational ideas. It was fed by many external influences, involving migrants and the skills and knowledge they brought, cultural values and affinities, economic circumstances, various borrowings, and 'supportive' technologies. Connecting to a scene may have happened through a common endeavour: acquaintances assembled for an industrial purpose, or an outcome of some co-venture. Networks overlapped, came and went. Still, 'network' is a convenient way to frame the connections people made, and that is what it will be called here for the purpose of discussion and further dissection.

Britain's 'national story' has to be addressed. The idea of the 'scene' stands at odds with the 'great inventor' myth still holding traction, even with some historians.[1] The creed was planted in Thomas Carlyle's era, *c.* 1840. Carlyle's view was that 'the history of the world is but the biography of great men'. These were sole originators, 'these great ones; the modellers, patterns, and in a wide sense creators, of whatsoever the general mass of men contrived to do or to attain'.[2] Carlyle was very influential. His ideas found their time, speaking to the new breed of business and political leader. By then, in commerce and manufacture, the eighteenth century's great advances were moving beyond memory. Even within one lifetime, personal recollections blurred. The discrepancies within popular Victorian versions of industrial history were, however, more than memory lapses. There was a desire to satisfy a new corporate world, the self-justification and self-reinvention Dickens characterized in *Hard Times* (1854),

---

1   See above, pp. 1–2.
2   Carlyle, *On Heroes*, Lecture 1, 3–4.

in the person of Bounderby, the self-made man who was no such thing. The Victorian reading public was also enthralled by Samuel Smiles, who from the 1850s fleshed out Carlyle's assertions with best-selling biographies of famous and successful engineers and 'men of invention and industry'. While presenting useful accounts of significant innovators, Smiles's works, and others in the mould, have limited context and unmistakable purpose, a partisan endorsement of individualism and hard work laced with Victorian sentimentality.[3]

So the myths are powerful.[4] 'Heroes of invention' narratives, though, the binary 'inventor/innovator' view, stand in the way of understanding. The authentic experiences described in previous chapters show alternative possibilities: that a multitude of lesser-known people were central to creating new industries and technologies. To discount that evidence means discarding valuable lessons which might inform our own times. Industrialization's early phases suggest novel ways to develop, organize, and regenerate.

Scenius

Eighteenth-century industrial development – and this applies whether the innovation was on a great or a small scale – can be understood as a creative process. In the performing arts, or in team sports, it is accepted that supportive environments – individuals as collaborators, with mutual learning and shared experiences – generate outcomes that amount to more than the sum of the parts. Arsène Wenger, a successful and unusually thoughtful football manager, has described what can be conjured when a team works in harmony:

> My never-ending struggle in this business is to release what is beautiful in man ... There is a kind of magic when men unite their energies to express a common idea.[5]

The musician and artist Brian Eno presented a concept for understanding collaborative creativity which he calls Scenius. Following an interest in Kandinsky, Eno discovered that the innovative Russian painter had been nurtured and encouraged within a budding art scene among other developing artists. Many of that set, though now less well known than Kandinsky, went on to great success. The group enjoyed practical support from patrons and galleries, who offered premises and competed for the artists' work. Above all, there was creative connection and advancement.

---

3 Matthew, 'Samuel Smiles', DNB.
4 Richard Evans, quoted by A. von Tunzelmann, 'Do we really need a national story?', *Guardian*, 11 Dec. 2021.
5 Interviewed for *L'Équipe*, 9 Oct. 2015. Here and below, references to 'men' reflect times and circumstances in which statements were made, but apply regardless of gender.

Eno advocates that young people 'learn to imagine' and that talents of whole communities be channelled into such a network, an ecosystem rather than 'ego system':

> And you get something that is actually an ecosystem. Now this thing about ecosystems is that it's impossible to tell what the important parts are. It's not a hierarchy, you know. We're used to thinking of things that are arranged in levels like that, with the important things at the top and the less important things at the bottom. Ecosystems aren't like that. They're richly interconnected and they're co-dependent in many, many ways. And if you take one thing out of the ecosystem, you can get a collapse in quite a different place. They're constantly rebalancing. And I feel that culture is like that.

The Russian artists, British pop culture from the 1960s, arguably even Renaissance Florence – all potentially classify as Scenius. Individuals may articulate ideas, but the influence alongside them is 'a whole scene of people who were supporting each other, looking at each other's work, copying from each other, stealing ideas, and contributing ideas'.[6]

This is not to deny that genius exists, rather an argument to place it in context, and to award fair credit for how it was generated. The most brilliant innovator is fostered within a wider culture, and builds on a platform of 'what already is'. This point, that community and context were fundamental to innovation, appears incontrovertible. From that starting position, many questions follow.

The web of connection

While the skill was the thing – and 'skill' embraces manual and intellectual expertise of many varieties and in numerous industries – networking made the difference. Through connection came real dynamism. But how did this process unfold? Most social connection is, and was then, entirely unremarkable. Yet something in this situation coalesced, and (whether or not they fully saw it) people were drawn into the endeavour, one which is now seen as very remarkable indeed. They recognized some possibility, an opportunity, and joined with others to achieve it.

---

6 Brian Eno, BBC Music John Peel lecture, 27 Sept. 2015: http://downloads.bbc.co.uk/6music/johnpeellecture/brian-eno-john-peel-lecture.pdf ; also https://www.philosophyforlife.org/blog/bowies-genius-and-enos-scenius; https://austinkleon.com/2017/05/12/scenius/. (Sites accessed August 2021. Likewise all other web sources.) Andrew Haldane, chief executive of the Royal Society of Arts, met with scorn when claiming Florence as a hive of transformational forces, alongside British industrialization: foreword to the government white paper 'Levelling up the United Kingdom', pub. 2 Feb. 2022.

The individuals introduced in previous chapters were many and varied, in background, skills and expertise, resources and interests. What they held in common is their role – sufficiently consequential to have been somewhere recorded – in the bigger picture of eighteenth-century innovation.

They appear here, in this account, because for some reason their work mattered. Numbers of them, by connecting with others, injected momentum into solving the problems in front of them. They are not a chance selection, but reflect activities described in previous chapters, things identified as key to the industrial upsurge. These figures represent the iron trade, textiles, civil engineering, collieries, commerce and finance, and also the new machine-making trade. There are elements here of industrially minded science, scientifically inclined industrialists, and educationalists in various forms.

The list appearing in the appendix, below, contains about ninety individuals and references some further associates. It ranges from lowly manual workers to people of great influence, though most are somewhere in between, often artisans, professionals, or gentlemen.[7] This is a heavyweight assemblage. Certain of them are seen as icons of British industrialization, while others, whether or not well known, are recognized as having made some notable contribution. This set is selective, embodying certain key pursuits. In doing so it forms a significant section of who was who in industrializing Britain.

These individuals have already been identified in earlier chapters as active and creative. Certain mutual links are obvious, while other important ones – exposed through more targeted research – enhance our knowledge of how such people interconnected. The investigation also clarified the nature of their industrial and personal relationships, and how significant those were to industrial advance.

To frame this discussion, two points should be kept in mind. First, 'innovation' here is seen as more than industrial hardware: it embraces supportative technologies, working techniques, developments in transport and other infrastructure, advances in factory design, industrial systems and arrangements, new sources of finance, and more. Second, cause and effect must be demonstrated: how precisely did connection translate into industrial change? Can collective endeavour be substantiated – where and what was the 'scene'? Where is the evidence of information and ideas arriving, and fuelling inventiveness, in workshops, factories, and ironworks?

What follows is inevitably discursive, starting with a summary of the main features of the ninety or so making up the list. The remainder of this chapter discusses how Scotland did it differently, for there was a significant and instructive contrast between its own experience of industrialization and that in the rest of Britain. Finally, the outcomes and the endurance of innovating communities are considered.

---

7 Brief biographies of the individuals appear below in the appendix, 'Webs of connection'.

The findings about those ninety individuals are in many ways unsurprising. The nature of their original occupation was generally determined by the family's status. As this is a group of individuals highlighted as exemplars of change, it is not a shock to find that many changed their work and social standing over the course of their lives. Social class was much less a determinant of how much they travelled. Some passed their whole lives in one location, while others moved considerable distances, including overseas, to live and work. But the overriding picture is the opposite of insular narrow-mindedness.

Education was closely associated with family status, and sometimes also with religious belief. Nonconformists were excluded from English universities before the 1820s, and the better-off might have chosen instead to school their sons in Unitarian dissenting academies, or Quaker boarding schools, possibly followed by a Scottish university. For Anglicans, a local grammar school was often the best option, with additional mathematical or other tutoring available to boys of ability, or destined for the family business. Sons of artisans could attend grammar school, though most from that class learned only the basics, usually at a dame school, to read, write, and understand arithmetic. University was expensive, and mostly unaffordable.

Family connections, especially through marriage, were an important social glue, the advantage to a family business coming through financial, managerial, or technological support – or introducing a son-in-law where there was no male heir. At risk of exclusion, Quakers were obliged to marry within the sect, but could not wed first cousins.[8] Unitarians tended to choose spouses from their own denomination. But there was no monopoly on strategic marriage, and clearly some opportunism at work. The Spencers' rise to supremacy in the south Yorkshire ironmaking combines was eased, it appears, by tactical alliances. But how different was their cynicism, if that is what it was, from the similarly unromantic behaviour in the marriage market by other wealthy families?[9]

Membership of local philosophical societies was a point of connection in educated circles. No doubt this widened acquaintances, and brought people together in a common interest. But many would already have known each other, through existing business relationships, church or family, or other means.[10] The same may be said for Freemasons, and members of trade associations and professional bodies. Valuable new contacts emerged through Royal Society fellowship or connection to the Society of Arts, but those groups were intellectually exclusive and closed to all but a select few.

Among the ninety, some blatant networking was evident – cultivating a set of people for their potential usefulness. Others seem to have had few associates beyond their immediate horizons. This may have suited their own circumstances

---

8   Above, pp. 165–6.
9   See above, pp. 153–7.
10 See above, pp. 114–29.

and needs, just as limiting travel was not necessarily a bad thing. Arkwright was a serial 'cultivator', sensible in his own situation. Samuel Garbett, the Birmingham businessman with interests in chemicals and iron, was a networker extraordinaire, in Scotland and England. But so too was Josiah Wedgwood, who as a potter might have appeared marginal to much of what was happening around him in the west Midlands. Yet far from being an outlier, Wedgwood knew everyone, or so it seemed. An early brush with smallpox affected his health and his ability to throw pots. His talents turned instead to the overview, to management, developing integrated factories, bringing in useful technologies and transport links – canals, of course, were a godsend for pottery exports. Wedgwood was also a marketing genius, and Etruria in Stoke-on-Trent became a marvel of the age. From his perspective too, networking made sense. But it was not the only route. A towering figure, the consummate innovator William Reynolds of Coalbrookdale, well resourced and self-assured, was apparently self-contained, or as much as that could be possible. In many respects John Smeaton was similar in following his own vision, a civil engineer with a stream of brilliant concepts, and not hostage to geography, any specific industry, or the need to earn.

The period covered by this group of ninety is extensive, embracing several active during the English Civil War. That long view helps gain some perspective on the early spread of inter-regional links in ironmaking. The Dudley–Yarranton–Crowley association arising from the west Midlands is intriguing, carrying outwards to Cheshire, south Yorkshire, Cumberland and Tyneside, and other places, and later involving Hayford, Spencer, and Bertram. That group was key in establishing connections across northern England, and before 1700. While their trading was initially limited, it created pathways which newer industries – and water transport – picked up on.[11]

The routes were not only physical. Interpersonal ties could lead specialist workers to employment elsewhere, or inform where goods or services that were unavailable locally might be sourced. A trade might build on the strength of such tentative, risky, approaches between regions. The Brandlings of Newcastle, having acquired through marriage the Middleton estate in Leeds, introduced smiths and colliery viewers from their Tyneside base to extend the colliery and build its famous waggon-way in the 1750s. By the 1790s the Walkers of Rotherham were serving the needs of collieries and other industries across the whole of northern England, including casting and transporting the components of Sunderland's spectacular Wearmouth bridge in 1795–6. With such cross-regional transactions came cross-fertilization of experience and ideas.

Not all exchanges were so consequential. But much of the evidence already presented indicates significant commercial transaction between regions.

---

11 See above, pp. 21–2.

Industries connected, they innovated, and this process intensified across much of eighteenth-century Britain.

A galaxy of Scotsmen

Scotland's role in British industrialization was critical. Its unique situation arose from political circumstances around the creation of the United Kingdom in 1707. Afterwards, institutions emerged in Scotland which were open to engaging with both industry and learning. A longer-term consequence was that influential and well-connected Scots were brought into pivotal positions, on the south as well as north of the border.

The Clows described this as 'Scottish leadership in the English industrial revolution'. They saw the smaller nation as progressing from 'a state of comparative obscurity to become one of the intellectual foci of Europe'. Overcoming the disadvantages of Scotland's challenging topography provided strong motivation. So, for instance, chemistry's potential to improve poor soil was recognized. Alchemy had enjoyed royal support, from James IV (d. 1513) and his successors, and laid the grounds for a culture of experimentation. There was no gold, but here were seeds from which 'the Scottish chemical efflorescence expanded'.[12]

Certainly Scotland was not an underdeveloped backwater before the union. The later initiatives built on 'technical accomplishments ... of a high order' in previous centuries, especially in Highlands mining and iron-smelting. 'A galaxy of Scotsmen gave scientific direction to the industrial revolution', said the Clows – referring in particular to chemicals and ironmaking.[13]

The Clows make only passing reference to the Darien catastrophe. This, the Company of Scotland's ill-considered scheme to generate trade with Central America by establishing a colony in Panama, led to an investment mania. Its failure, 1698–1701, was a calamity, with great loss of life for colonists, and a devastating financial shock to the small nation. Rescue came at a cost: with the Act of Union in 1707, the Scots forfeited their parliament.[14] But investors were compensated, and a measure of institutional support was offered to address Scotland's industrial and agricultural difficulties. An annual £2,000 was allocated for seven years to promote manufactures and fishing. The fund accumulated, and for a time went unspent.

---

12 Clow and Clow, *Chemical Revolution*, ch. XXV, esp. 582–3.
13 Clow and Clow, *Chemical Revolution*, xi–xiv, 21, 41; National Records of Scotland, NG1.
14 See Hamilton, 'Darien investors and colonists (act. 1695–1707)', DNB.

In 1723 came the Scottish Society of Improvers in the Knowledge of Agriculture, developing an interest in soil quality and agricultural improvement.[15] Then, most significantly, the Board of Trustees for Fishing, Manufactures and Improvements was established in 1727. Funded with £6,000 a year, it aimed to bring scientific insight to bear on industrial and agrarian problems, and explicitly encouraged the linen industry. Significant advances nurtured by the Board, in chemicals and ironmaking, reflect the preoccupations of Scotland's university-based scientists, and particularly the expertise on hand in Glasgow and Edinburgh. This happy congruence of industrial need with the interests of Scottish scientists came with financial incentives from the Board which supplemented the generally low pay of academics. It can only have helped propel scientists towards chemistry.[16]

The Board was awarded further financial support, some ring-fenced for the linen industry. An initial focus on linen-bleaching was soon extended into soap and potash production, dyestuffs, sulphuric acid, and other chemicals. Scientists advising the Board, most prominently Joseph Black, suggested ways to improve traditional working practices. From the mid-century into the 1790s, Black and his one-time colleague at the University of Glasgow, William Cullen, supervised the Board's relationship with industry. With other men of science, they judged applications for premiums, and analysed and tested proposals. For instance, in 1783 Black, with a friend, the geologist James Hutton (by trade a chemical manufacturer), represented the Board of Manufactures in assessing a new design of furnace to produce pig iron at a Leith foundry. Their conclusion was unfavourable, but they took considerable trouble in reaching that decision.[17]

The Board, then, acted as a high-level forum through which science could advise industry. Equally, practical lessons were taken which informed scientific understanding. Black was part of an Edinburgh and Glasgow network of experimentalists and industrialists associated with the Board, who also met socially alongside invited visitors. The public intervention embodied by the Board proved an enduring structure through which Scotland gained some unique and significant advantage. Working with the Scottish universities, the Board's achievement was more than technological – it drove economic development and helped establish networks of innovators.

So there was a Scottish 'scene', and its synergies of industrial and scientific interests are verifiable. Nothing quite like it is evident in England, nor any equivalent public policy initiative. There, any ripples of interest in science's industrial potential were piecemeal, and the English universities far removed,

---

15 For the Society of Improvers in the Knowledge of Agriculture, 1723–46, see Bonnyman, 'Agrarian Patriotism and the Landed Interest'.
16 Anderson, 'Industry and Academe in Scotland', 339.
17 Clow and Clow, *Chemical Revolution*, 41, 67, 136; Anderson, 'Industry and Academe in Scotland', 342–3 and *passim*; 'Joseph Black', DNB.

in locality as well as in spirit. William Lewis, it should be recalled, one of a very few experimental chemists interested in industrial possibilities, chose to work outside an institution. The Scottish project was meanwhile attracting men of similar mind and great ability to turn their focus to very practical matters. Black, in particular, took chemistry out of the shadow of medicine, and then demonstrated its potential for industrial purposes. Closely identified with the Board of Trustees, Joseph Black carved a role advising Scottish industry. He became something of an industrialist himself – a path later followed by others whose grounding had been in science.[18]

What of the Clows' claim of 'Scottish leadership in the English industrial revolution'? Black, it might be remembered, worked alongside James Watt and facilitated his experiments. But it was evidently John Roebuck, Carron Company partner and brilliant industrial chemist, who introduced Watt to Boulton, an act which ultimately dispatched steam-engine manufacture to Birmingham. This was very much how Scottish influence was carried, in human form, by migrants and visitors. Roebuck, though, was not Scottish. Better travelled than most, he came from Sheffield, attended the Unitarian academy in Northampton, then the universities of Edinburgh and Leiden, and worked in Birmingham and at Prestonpans before Carron. His career illustrates directions and networks through which knowledge moved. Roebuck's best industrial work was in Scotland, and he is considered a pioneer in commercializing Scottish science.[19]

Carriers of knowledge and experience, then, were not all Scots-born, and benefits flowed in many directions. But the Scottish education system was instrumental in this process of creating connection and shaping science to industrial needs. It was especially so for chemistry, as it related to bleach, dyestuffs, and ironmaking. Scottish universities were a magnet for the nonconformists (broadly defined) barred from English higher education. Scotland offered learning of high quality and practical focus, and there was no direct equivalent to this in England. This distinctive approach was also true of schools. Scottish children's elementary schooling evidently outshone the English norm. That too had effects south of the border, in the quality of young Scottish migrants who later proved themselves in civil or mechanical engineering: William Murdock of Boulton & Watt; William and Peter Fairbairn; and M'Connel & Kennedy, who travelled from Kirkcudbright to be apprentice machine-makers in Lancashire.

Many alumni of Scottish universities also stamped their mark in England – including Rennie, Roebuck, and perhaps Watt, though he was not strictly an alumnus of Glasgow. Others returning home to England similarly made waves: some of the Manchester Lit. & Phil. bedrock (Percival, Henry, Watt jun.)

18 Marsden, *Watt's Perfect Engine*, 21–2; Anderson, 'Joseph Black', DNB.
19 Clow and Clow, *Chemical Revolution*, 93–4, 587, 592, 594; Campbell, 'John Roebuck (1718–1794)', DNB; Norris, 'Struggle for Carron'.

and the likes of Erasmus Darwin and Joseph Dawson. Oxbridge, with few industrial links, could not serve their needs, socially or educationally. Chemistry classes were sometimes available at Oxford and Cambridge, but casually and unexamined. The subject emerged more distinctly in the universities of Edinburgh (from 1726) and Glasgow (1747), and in Aberdeen's colleges from 1793. Edinburgh also benefited from its close relationship, formalized in the 1710s, with the religiously tolerant University of Leiden (founded 1574). Certain Edinburgh medical students went on to Leiden to complete an MD degree, including John Roebuck, Thomas Percival, and William Brownrigg, Charles Wood's Cumberland associate who was similarly interested in metallurgy and chemical elements. Many who taught medicine at Edinburgh, and at Glasgow and Aberdeen, were graduates of Leiden. The influence of Leiden's star-studded faculty was pervasive, and enduring. John Dalton (mainly self-educated) was inspired in his chemical endeavours by the writings of the brilliant Leiden professor Herman Boerhaave (d. 1738).[20]

The chair established at Edinburgh in 1713 was in chemistry and medicine, disciplines which continued to run side by side. To train as a physician was a recognized route into science, and a shift towards chemistry does not necessarily indicate an abandoned medical career. Science may always have been the objective. Some in medical practice (which paid) continued to pursue that in order to support their scientific work (often unpaid, and expensive to fund). This is comparable to academics using university pay and facilities to sustain research, or science-minded ironmasters (there were a few) using business resources to run trials. Some distinguished researchers who viewed science as a vocation, a higher calling, were obliged to seek a living in associated work. John Dalton was one. Money was not an issue for Joseph Black, though even he (probably for social reasons) continued a very small and exclusive medical practice, including being appointed principal physician in Scotland to George III.[21]

Edinburgh, near to Scotland's industrial belt, was convenient for Carron and Prestonpans – all these on the Firth of Forth – and other sites developing after the mid-century. Significant links were built between certain academics and businesses, though roles were clearly delineated. Industrialists, says Anderson, did not teach, while almost all chemistry education was delivered by men with medical degrees. Early in the century, Edinburgh academics were indeed active in commercial production, but only as far as it related to medicine. Into the 1740s, unsalaried and with student fees insufficient to support them, some sold remedies to local pharmacies. A generation later, Thomas Henry,

---

20   Clow and Clow, *Chemical Revolution*, 583–5, 602–3, 609, 39, and *passim*; Anderson, 'Industry and Academe in Scotland', 338–9.
21   Anderson, 'Joseph Black (1728–99)', DNB. Fortunately, the king did not venture so far north.

the pharmacist who taught chemistry at Manchester Academy, found a more robust source of funds for research by setting up the profitable magnesia factory financing his and his son's chemical investigations.[22]

Throughout the eighteenth century, though, even as industrial connections with England were strengthening, Scotland did things differently. It built outwards from its strong, university-oriented model of sciences that were relevant to industry. It is remarkable that in 1814 this small nation surpassed the rest of the world per capita in 'intelligent practical chemists'. That was the claim, and it sounds entirely credible.[23]

How had it come about? One influential development, uniquely Scottish, was the chemical societies attached to its universities. The first appeared in the 1780s, and there were four of these societies by 1800. The fifth was in Philadelphia, founded (almost inevitably) by Edinburgh alumni. Alongside this came an upsurge in extramural teaching in Scotland. John Anderson founded his own institution in Glasgow, a model which George Birkbeck took to London, and the concept was later advanced through Mechanics' Institutes. But in this Scotland was some way in advance of England, in chronology as well as in the quality and practical application of the syllabus. In Aberdeen this new style of education launched in 1783, with a focus on chemistry that was applied to arts, manufactures, and agriculture. From there followed conscious attempts by certain teachers to distance chemistry from medicine.[24] Chemistry was promoted as industrially relevant, a subject in itself, and was becoming more accessible. News of its possibilities opened up to people previously lacking an entrée to that world.

The pragmatic approach is reflected in another Scottish initiative of 1783. Glasgow boasted the first Chamber of Commerce in the UK – the second would follow quickly, in Edinburgh in 1785. The Glasgow pioneer was highly active, concentrating on the local cotton industry, on developing dyestuffs and calico printing, and on chemicals in general.[25] Yet Glasgow's first philosophical society did not appear until 1802 – an early speaker there was the younger James Watt. Indeed, the Watts bridged Birmingham, Manchester, and Glasgow. When Watt the elder departed Glasgow for Soho in 1774, he retained strong links. Enduring friends, fellow members of the Literary Society of Glasgow (founded 1753), included Black and Anderson, and Watt stayed close to Black thereafter. The pair were simultaneously experimenting on heat, and Black first presented his findings on heat transfer, resulting in the concepts of 'latent heat' and 'specific heat', to the Literary Society in 1762. Here was a forum which

22 Anderson, 'Industry and Academe in Scotland', 335–6. For Henry, see ch. 5.
23 Clow and Clow, *Chemical Revolution*, 599, citing Sir John Sinclair's *General Report of the Agricultural State and Political Circumstances of Scotland* (1814).
24 Clow and Clow, *Chemical Revolution*, 597–609.
25 Clow and Clow, *Chemical Revolution*, 597–9.

in any modern sense was more than literary. When Watt became preoccupied with bleaching and then gases, from the late 1780s, he experimented on the Glasgow linen fields of his father-in-law James McGrigor, and followed those same interests to the Manchester Lit. & Phil. and its experimentalists.[26] Watt's approach reflected that of Scottish science during his Glasgow career, the focus on practical purpose. He carried that with him ever afterwards.[27]

In Scotland overall, many men of science (not all of them academics) were enlightened by the workings of industry, sometimes to the extent of using industrial premises as places of experiment and trial. The promise that science held for business was also recognized, especially in agriculture, textiles, and ironworking. Even in those sectors, though, not every Scottish industrialist understood, or was convinced, or chose to pick up on that potential.

Carron, itself indebted to the second Abraham Darby's organizational model, acted as a primer, a template for further ironworks developed across central Scotland. These were largely around Glasgow and involved former Carron employees.[28] Carron itself, though, was soon in trouble. John Roebuck's technological ambition loaded the company with unsustainable debt, which eventually sank Samuel Garbett and led Charles Gascoigne, Garbett's son-in-law and Carron manager from 1769, to flee to Russia in 1786.[29] During Gascoigne's time, most of the original projectors were squeezed out. Earlier, from its building in 1759–60, Carron had been an influential hub. Most involved in its launch were English, some having been educated, or holding other interests, in Scotland. At a time when – at the most – 17 coke-blast furnaces were at work in Britain, Carron set out to build 4 blast furnaces and 4 air furnaces on its site. Its skilled workforce too was mainly from England, including Thomas Cranage of Shropshire. John Roebuck took a significant responsibility for everything technical. When the blast was insufficient, John Smeaton was commissioned to design a blowing cylinder. Later Smeaton advised on water power for the boring mills and other infrastructure matters. A marker was laid down by Carron: it represented a new kind of operation, drawing on the finest available expertise, theoretical and practical.[30]

---

26   Clow and Clow, *Chemical Revolution*, 598; Anderson, 'Joseph Black (1728–99)', DNB; Tann, 'James Watt (1736–1819)', DNB.
27   Boulton & Watt's defence of their intellectual property, and abandonment of improvements which could not be patented, have been discussed: pp. 126–8, 139.
28   For Carron's wider influence in Scotland, and its effect on Scottish pig iron production, see Clow and Clow, *Chemical Revolution*, ch. XVI, esp. 341–6, 360.
29   Campbell, 'John Roebuck', DNB; Chrimes, 'Charles Gascoigne', BDCE; Norris, 'Struggle for Carron'.
30   Clow and Clow, *Chemical Revolution*, 333–41; Hayman, 'Shropshire wrought-iron industry', 55; Schubert, *History of the British Iron and Steel Industry*, 332–3; Robinson, 'Charles Gascoigne', DNB.

Carron's importance extended beyond iron production and 'hard' technology. It was a hub that generated and nourished personal connection, just as did Edinburgh's medical school. The ripples of influence took no account of borders, spreading across Britain's industrial districts and beyond. Significant names from the list of ninety or so, representing science and iron, civil and mechanical engineering, and commerce, had a direct and enduring connection with Carron: Black, Cranage, Garbett, Gascoigne, Roebuck, and Smeaton. Henry Cort and Samuel Owen were linked. Many of these ties are clear, but some are obscure and not yet fully explained. Why, for instance, did the Leeds machine-maker Maclea, Scottish son-in-law of Matthew Murray, bear the forenames Charles Gascoigne?

But even in this encouraging milieu, even in the 1790s and later, when it came to embracing chemistry there was ambivalence in the Scottish iron industry. David Mushet, from a family of ironmasters, joined the Clyde Iron Works, an offshoot of Carron, as accountant in 1791–2. In his spare time he investigated iron-smelting and metallurgy, and was briefly allowed access to experiment in the works. From 1794, Mushet had to work with his own small furnace elsewhere. He seems to have been dismissed from the Clyde works in 1800, and with partners then established Calder Iron Works, near Airdrie. He was responsible in 1800–1 for discovering the blackband ironstone which underpinned a great expansion in Scottish ironmaking; patented an important development in direct steel-making in crucibles (1800); and zealously laid the foundations of a technical literature in his industry. Mushet is seen as the most important investigator of iron and steel technology in Britain in his time. In 1805 he left the Calder partnership, and Scotland, to live thereafter in England.[31]

Mushet's experience echoes that of his friend, the scientifically minded Joseph Dawson of Low Moor, in the resistance his ideas met among other ironmasters. Dawson, though, as managing partner, enjoyed much more independence in setting his own company's path. The unanswered question here is why would it be, in instances where science appeared to offer industrial solutions, that these were not welcomed by some industrialists? How had Mushet's frustrating situation arisen, the uneasy relationships, the loss to Scotland of this great intellectual and technological asset?

The reality for ironmasters in the 1790s was that the transformation which puddling enabled was double-edged. Within two decades the system had been adopted by almost all ironworks. Puddling was not optional, but quickly became imperative. To label the Clyde partners 'conservative masters' who objected to Mushet's progressive scientific approach may therefore be unfair.[32] The fact

---

31 Standing, 'David Mushet (1772–1847)', DNB; Clow and Clow, *Chemical Revolution*, 170, 342, 344, 351–3, 357. On the importance of Mushet's scientific publications, especially in the *Philosophical Magazine*, see Pacey, 'Emerging from the Museum'.
32 Clow and Clow, *Chemical Revolution*, 342.

is that they, like directors of nearly every sizeable ironworks, were forced to confront other priorities during the decade Mushet spent in their employ. Puddling incorporated many existing processes but also required changes. The new system used some new technologies and different manual techniques, meaning additional training and re-equipping, finding capital investment, and dealing with critical skill shortages. It was manageable, for ironworkers had always readjusted to changing technology, but this was an extraordinary time. Such was the demand for iron, the need to compete on prices and to build new capacity, that no company could afford to be left behind. By 1815, British pig-iron production had risen from 90,000 tons in 1790 to almost 400,000; bar-iron, the output 32,000 tons in 1788, had reached 150,000 tons. Mushet's discovery of the blackband ironstone may have been decisive in empowering Scotland to respond and increase its output relative to the rest of Britain.[33]

For the Clyde Iron Works owners, dispensing with Mushet's services could have been quite rational. Research is expensive, and Mushet, even funding his own investigations, was a distraction from other pressing matters. His relevance to the business was perhaps not obvious. Innovation, as Murdock's experience with Boulton & Watt shows, is not always welcome. The course of Mushet's subsequent career suggests that his commercial acumen did not match his brilliance in other spheres, and that could have been a factor in his departure from both Clyde and Calder.

Innovating communities

Community took many forms. It could interlock or overlap, be exclusive or open, be industrial, commercial, cultural, or some blend of those. Where community operated at innovation's sharp end, from the mid-eighteenth century its connections shaped a new era in the iron and machine trades.

Business premises, a company's outward expression, show how operations were accommodated and how they evolved.[34] They can also suggest less tangible features: associations with suppliers and customers, other local contexts, and something of the industrial ethos. The habits, the patterns, of earlier ironworking had been transported into new locations, where they adapted to new needs, technical and managerial.

Late in the eighteenth century, forge tradition had a clear influence on the physical form of machine-making as it emerged. A 'best possible' system, a dispersed factory of workshops, interlocked into one production system. The new settings generated networks of subcontractors, most clearly shown within the orbit of Richard Hattersley in Keighley. Flexibility suited the tight financial and technical constraints of the start-up industry, and that situation

---

33 Clow and Clow, *Chemical Revolution*, 351; Hyde, *Technological Change*, 106–12.
34 Thanks to Colum Giles for suggesting more attention be paid to premises.

continued in machine-making for some time after textiles had become largely factory-based. Even after the 1820s, as machine-makers integrated new systems into their works, for many engineers in supplementary and specialist roles a workshop with a small steam engine was still sufficient.[35]

Before then, what do premises say about the industrial cultures at the vanguard of innovation and transformation? Extremes of scale were a feature of ironworking, and that was true throughout the period. The size of a works was determined by its products and range of operations – whether furnaces, forges, or both. Highly capitalized enterprises bundled their activity into groups to spread risk (the Spencer and Foley combines). Extensive sites may be components of even larger ventures (Crowley, Darby). At the opposite extreme were small and transient charcoal-burning furnaces, placed close to a finite source of fuel. Many of the forges in operation were no more than one hearth working part-time in a backyard smithy.

Where can 'Scenius' be found in this? It emerged out of the convivial working which had long been practised across ironmaking. Those older templates were then adopted, and adapted, when machine-makers came to devise their own arrangements. Hayman shows that Midlands forging and iron-goods production depended on collaboration: because fining and hammering demanded different skills, it was impossible for one master craftsman to manage the whole. Specialisms passed down through generations, for the trade was usually based on family units working in leased or rented workshops, with a single hearth and bellows, and this was commonly combined with farming.[36] Such were the patterns across regions, and also in continental Europe, dictated by the demands and limitations of forge work and by other local needs. The Wortley neighbourhoods' small rural nail-making and wire-drawing forges were similar, and so too Sheffield's 'little masters', the independent tradesmen characteristic of cutlery, tool-making, and other highly focused and skilled trades.[37] Those examples display varied specialisms and skill levels, but in common were a degree of collaboration and the family and commercial networks which sustained them. Meeting, linking, and sharing information were necessary to business; in that, there rested conditions to spark a creative energy. A small specialist workshop could be, and quite often was, a kernel of innovation.

The iron trade, though, also worked on far larger platforms. Some operations must be near to minerals, fuel, water power, and transport, and be adapted physically to specific sites. A furnace might require round-the-clock attention,

---

35 Cookson, *Age of Machinery*, 135–41, 232–5. Matthew Murray was unusual in having his own foundry and a more regimented organization, from *c.* 1800, though much of his site was dedicated to producing steam engines. The Drabbles, in Jubb's former premises, were more typical machine-makers, driving machine tools in 1803 with a 6-h.p. engine: Science Museum Lib., Goodrich mss, GE2 (with thanks to Edward Sargent).
36 Hayman, 'Shropshire wrought-iron industry', 147.
37 Hey, *Fiery Blades*, 2, 102, 287; Kelham Island Museum: https://www.sheffieldmuseums.org.uk/whats-on/little-mesters-street/.

in a place close to iron ore and charcoal. The result (similar to arrangements commonly found in coal-mining) was semi-rural colonies containing workers' housing.

The Swedish *bruk* existed for centuries before these large projects were first contemplated in Britain. Did Sweden provide a model? The meanings of *bruk* include ironworks, forge, and practice. Evans and Rydén define it as an ironmaking estate with at least one forge and perhaps a blast furnace, located in abundant charcoal woodland. Its workforce was stable, with generations living on the site and working the forge day and night. In significant features, however, it differed from British experience. The careful management and monitoring of such ventures in Sweden served that country's uniquely strategic overview of its iron production. In the eighteenth century, too, the trend was towards *bruk* workers ceding much of their autonomy.[38] British large-scale ironworking was very limited before 1750, the topography and access to charcoal ill-suited to *bruk*-style arrangement. Early works of all sizes were individualistic, their design reflecting a developer's purpose and vision.

Coalbrookdale, distant (in contemporary terms) from other iron producers, represents a slow evolution working on difficult terrain. Its locational advantage was in access to good minerals, and the steep descent towards the River Severn. From 1658, half a century before Darby, some small and tentative works were created. Building over time, the Darbys colonized all possible areas in the dale, and moved into the Severn Valley as processes grew more complex and older plant deteriorated. Natural resources were carefully managed to best effect, against the inconveniences of gradients, constricted sites, and the limits of water for power and transport.[39] Houses, cottages, and workshops were slotted in where they would fit. After 1750 the family developed satellite sites and, latterly, Coalport. It was a spectacular vista, but not a package others would choose. In premises, as in other aspects of business, Darbys took their own course.

Ambrose Crowley's Swalwell Works on the Derwent, a southern tributary of the Tyne, appears to have been planned largely as a piece. On a relatively flat site, water from the Derwent was engineered to feed a slitting mill, steel furnace, large forge, and many small workshops set in neat rows. The inventory and site plan of *c.* 1728 show its vertical integration and high level of organization, culminating in a wide range of finished items. Streets of cottages suggest a settled community, though this was of quite recent construction. Housing was provided to attract a workforce from beyond the region to this out-of-the-way place. To Crowley, Swalwell was a cog within a very large wheel, designed to

---

38 Evans and Rydén, *Baltic Iron*, 86–8, 16–17; Jansson and Rydén, 'Improving Swedish Steelmaking', 517.
39 Hayman, *Ironmaking*, 133–9; Baggs, Cox, et al., *History of the County of Shropshire. XI. Telford*, 40–56.

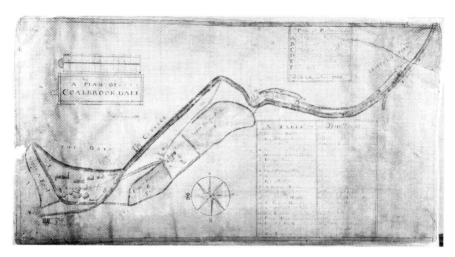

Plate 12. Thomas Slaughter's Plan of Coalbrookdale, 1753, shows complex systems of work and water management. On the steep descent from left to right are the main ironworks site, and some of the Darbys' houses which overlooked it; lower furnaces; upper forge; and the lower forge, almost on the riverside road alongside the Severn. Source: Ironbridge Gorge Museum Trust, 1971.26 – Slaughter Map.

meet contracts supplying the Royal Navy with 108 types of nail and dozens of other goods. It was intended to be controlled from afar, as demonstrated by the Law Book of the Crowley Ironworks, unique in its time with its roles and responsibilities, and penalties for transgressions.[40] The international division of labour noted by Schröder as an 'iron system' in the 1740s was shaping up under Crowley.[41]

The shop-floor scene, like the iron trade itself, grew in vibrancy and influence from the mid-century. Forge culture became powerful enough that some larger ironworks struggled to introduce under-managers and piece-rates. Skill and strength in combination gave forge-workers collective leverage to insist on direct employment. That the men themselves controlled entry to prized trades particularly frustrated owners, who railed against workers' 'ignorance and vile wickedness'. The slurs were hyperbolic, for in the workshops were men of great intelligence and understanding. Indeed, having master forgemen organize their own working groups suited managers. Shop-floor creativity was the source of much innovation – new ideas, and bringing others' concepts to

---

40 Evans and Rydén, *Baltic Iron*, 193–6; Price, 'Ambrose Crowley (1658–1713)', DNB; Flinn (ed.), *Law Book of the Crowley Ironworks*; T&WA, DX104/1, c. 1728.
41 Evans and Rydén, *Baltic Iron*, 12–18.

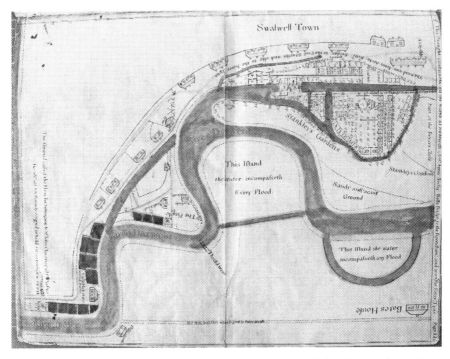

Plate 13. Plan of Crowley's Swalwell Works, *c.* 1728, which had been inherited on Ambrose Crowley's death in 1713 by his son John. Source: T&WAS DX104/1, 'A draught of Esq. Crowley's Works Beginning at The High Dam above Winlaton Mill and Ending at The River Tyne'.

work in different contexts. The charge that workers resisted innovation was false, says Evans, about south Wales, for 'informal craft skills underpinned formal technological advance'. Skills were tactile and sensory, absorbed within working environments.[42]

Hayman argues that skilled Midlands ironworkers had little to fear from innovation. Collaboration was fundamental to how the trade worked, and forgemen were relatively privileged. Their skill protected them: no matter how often a method was explained or demonstrated, it took empirical knowledge and long experience to repeat it. This explains why new practices were shown to visitors without qualm, even to iron-trade experts such as Alexander Chisholm, Simon Goodrich, and Angerstein himself. When Shropshire forgemen readily adapted to puddling, a departure from what they were accustomed to, they

---

42   Evans, *Labyrinth of Flames*, chs 4 and 5, esp. 75–80. South Wales masters objected to the customary 'immersion' and 'lore' which stood in place of indentured apprenticeship.

protected their own position. Recognizing puddling's immense promise, that it would become ubiquitous, they knew that the technique had to be embraced and applied to local circumstances. Skill conferred authority, for instance to insist on retaining certain older methods alongside the new. Stamping and potting was tweaked for specific purposes and products in the Shropshire industry, and charcoal still used as required to produce high-grade iron.[43] Here was the forgeman's skill: appraising technologies, assessing their effectiveness in different usages, adjusting to achieve a 'best possible' result. Hayman puts forward a compelling argument that technological change was a continuum, not sequential. Forge trades adjusted cautiously to new practices, while continuing with older technology – perhaps adapted – for as long as it offered something additional.[44]

Together, Hayman's Shropshire conclusions, and Evans' on Merthyr's forge culture, show how imperative is an understanding of the world of work. It is there that new technologies' origins are to be found and assessed. In the eighteenth century, and not only then, innovation in key sectors – notably iron-processing and machine technology – was largely a product of the workplace, the shop floor. In such settings rest vital insights into the process of innovation.

The forgemen's 'collaborative culture' is especially remarkable. Their talents, shaped by a unique working environment, resided in manual skills demanding both strength and sensitivity. Thus they judged iron before formal chemical knowledge existed, and made technical decisions to renew, adapt, or discriminate. That included adjusting to a new system, like stamping and potting, for what it could offer, while also appreciating that it was imperfect. Also clear is that some of the innovation was at a micro-level – an expedient to solve a very localized problem, or recruiting other expertise within a neighbourhood – for example, knowledge of steam power.[45] There were specific local needs, as well as different ways of working in different ironmaking regions. Even within one ironworks, distinctions are evident in how the forge and foundry operations were managed.

How innovative were the new kinds of ironworks, the ventures launched from the 1740s? Innovation, while common, took many forms. Only in part was it solidly technological. Even for factory processes, novelty might come from techniques and methods. Fresh products could be highly significant, as could organizational efficiencies and new commercial practices. There was no guarantee that companies known as technological pace-setters would sustain their success. New systems of managing work, dull as that sounds, could be

43 Hayman, 'Shropshire wrought-iron industry', 149, 116–17. Hayman shows that puddling was not a natural progression from the Cranage method.
44 Hayman, 'Shropshire wrought-iron industry', 108–13, 116–17, 149, 214–17.
45 Evans and Rydén, *Baltic Iron*, 16; Bruton, 'Shropshire Enlightenment', 114–15; Hayman, 'Shropshire wrought-iron industry', ch. 6; Evans, *Labyrinth of Flames*, esp. ch. 5.

transformational. Introducing method to the randomness of subcontracting and out-work had applications beyond ironworking. From Ambrose Crowley at Swalwell to the Walkers in Rotherham with their managing agents and departmentalized director functions, a trail leads towards Benjamin Gott's ambition for Leeds woollen-cloth manufacture in the 1790s, his gargantuan factory built to house old technologies, while a judicious mix of foreman-organized out-work busied itself elsewhere. And this meant networking, always networking.

From the forges of the new ironworks, and those of south Yorkshire nailmakers, came other threads of connection. Kirkstall Forge, earlier a part of the Spencer combines, was revived by Butler & Beecroft from 1779. A decade later, the company cultivated their market in manufactured goods, spades and shovels, industrial components, and items for Leeds's booming domestic demand. To this end, The Square was built, on a site between the river and Kirkstall's water-powered slitting mills and dam. The Square – actually a large rectangle – contained around a dozen inward-facing workshops, each with hearth and high chimney, some with integral cottages. Unlike other workers at Kirkstall, the occupiers of these workshops apparently signed up as subcontractors. It was almost a replication of Wortley's nail-making model, here gathered into neighbouring industrial units. The Kirkstall plant overall was outward-facing, maintaining strong connections back to Wortley ironworking, building new links across the industrial districts around Leeds, welcoming customers and the other visitors who streamed through the plant.[46]

In his brief stay at Kirkstall, Richard Hattersley absorbed this organizational model and, carrying elements of it to Keighley in 1789, also created his own business link back to the Forge. In Keighley he nurtured a hub producing textile-engineering components, the flyers, fluted rollers, and spindles which could not then be machine-made. Through constant repetition of a manual task, the component-makers could deliver just-about-acceptable measures of precision. Hattersley's closeness to his subcontractors, and to many Leeds customers who were part of the Wortley diaspora, some of them his relatives, established a rapid feedback loop.[47] This level of proximity across the new machine-making trade cemented connections, circulated technical knowledge, and sustained creativity. It was a mass of contingency, and it worked.

That all examples were somehow exceptional may be true. Workplaces differed in products, processes, and premises. Needs and resources varied across localities, while the context everywhere was one of flux. But more interesting are the congruities, across regions and even nations, the patterns that are evident

---

46   Butler, *History of Kirkstall Forge*, 11, 16–17, 32, 107, 80; WYAS Leeds, WYL76, Kirkstall Forge inventory, 1783. George Seed's detailed plan of Kirkstall Forge in 1844 (viewed at Commercial Estates Group Ltd offices, Kirkstall Forge, Nov. 2019) shows The Square partly occupied, by wheelwrights and for wood storage. It was demolished in 1847.
47   Cookson, *Age of Machinery*, 139–41.

CREATING A SCENE 191

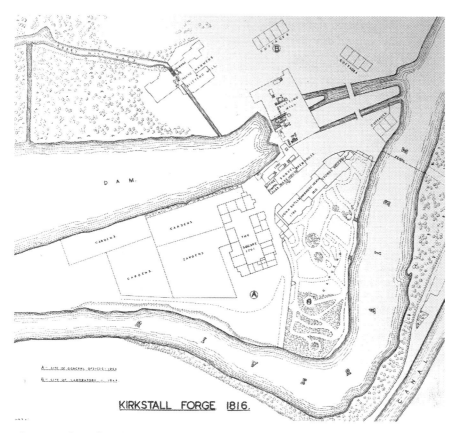

Plate 14. Plan of Kirkstall Forge in 1816. Source: Butler, *History of Kirkstall Forge*, between pp. 80–1.

even if local detail did not quite match. The iron trade, coherent and internationally connected, points the way.

Without doubt, ironworkers' mobility facilitated technological advance. The skills embodied in individuals were portable, moving easily within and beyond regions, transmitted through migration and other types of interaction. 'Skills' include immersion in a trade – working arrangements and systems, for example, as well as manual and intellectual aptitudes. So forge culture developed some consistency across the iron districts. The iron trade's international dimension was strongly influential, too, and for the long-term, particularly between Britain and Sweden. That association was key to both nations' advance. Its meaning was far beyond numbers on a balance sheet. Most obviously, here was the source of a substantial part of Britain's iron needs, and of Sweden's export market. And with this relationship fundamental to a global commodity chain, there was important reciprocation in knowledge, technique, and dialogue. This is why Angerstein and many other Swedish specialists in iron technologies were welcomed to Britain. Learning was a two-way street. Swedish ironmasters were open to adopting elements of British organization, including divisions of labour, the 'English method' recommended by Schröder.[48]

Ironworking consciously occupied a wider world which extended beyond other regions. Its web of influence brought awareness – including of failures in older systems, like the over-reliance on charcoal. It focused attention too on industrial pinch-points and unmet demand. Ironmaking, though embedded in tradition, was also (at least in certain companies) open to fresh thinking. There were mis-strides and false dawns, but the industry's capacity to produce solutions grew impressively. This was true even while the science of chemistry tried (with very limited success before 1800) to catch up.

Here, then, was a creative 'scene', encompassing local specialism and a reach which was international. The forge connections rested on a common terminology (or at least a common understanding of regional terminologies), and on familiar systems. Technological adjustments were accepted if they were promising; and there was readiness to modify working practices to accommodate something more radical. Thus, forgemen absorbed the Cranage system, recognizing its limitations but applying it where it could best serve, and subsequently went wholeheartedly for Cort's puddling process, but only once Crawshay's forgemen had made it workable.

And there, in Crawshay's intervention in Cort's technology and his company's consequent shop-floor trials, is shown a wider 'scene'. The workshop was an interface between forgemen and management, along with 'intermediate' occupations. The mix worked to great effect. Most influential were forgemen (hammer-men and finers) along with smiths and millwrights (with both those categories sometimes working as subcontractors, in Britain at least). The ironworks which increasingly stood out as innovative had a new model of ownership. It featured the active involvement of men of ideas and technical competence, though often with limited assets. Darby was a prototype, followed in the mid-century by the influential Cockshutts and Walkers, and later Dawson at Low Moor. The solution was to find partners who could invest, often merchants who knew the iron trade. Crawshay came to own Cyfarthfa through that route, buying into a promising concept and recruiting necessary skills.[49]

Science, if it appeared at all in mid-century ironworks, was generally there to learn. The approach of William Lewis, his scientific engagement with the iron trade evidently unique in Britain, is illustration. Later, in Manchester in the century's closing decades, a 'scene' of applied science emerged, after the Lit.

---

48 Thanks to Göran Rydén for views on Swedish–English transmission of 'learning'. See also Bruton, 'Shropshire Enlightenment', 115.
49 Rydén, *Production and Work*, ch. 3.

& Phil. had focused on dyeing, bleaching and gaslighting in close collaboration with cotton-industry interests, while Dalton worked to systematize chemistry. This too appears to have been exceptional in its time.[50]

The Wortley phenomenon – which happened contemporaneously, on the Yorkshire side of the Pennines – was as important to industrial innovation as was the Manchester science scene. This can be stated with confidence. To summarize 'Wortley phenomenon', it emerged from a fresh and inquiring attitude towards iron production at several south Yorkshire sites; and was the source of traffic in technological knowledge and skills from those places, most advantageously into Leeds and Keighley engineering. Those migrations mark practical advances from ironworking into machine-making, and were strongly supported in the following decades by the movement of further skilled workers and promising trainees.

These two remarkable instances, west and east of the Pennines, pursued different purposes. What they shared was scene-building, constructing a context for innovation. The setting was already well established, one of innovation and growth in textiles and other local manufacturing. The breakthroughs took matters to another stage, enabling new phases of industrial development. In Manchester there came chlorine bleaching and gas plant; the Wortley-to-Leeds engineering migrants delivered iron-framed machinery and a highly accomplished component industry, using hand skills for still-unmechanized processes. In both instances, direction was established, relevant experience assembled, specific problems addressed, initiative taken, wider social networks called upon for specialist support. Thus was created a new scene, based largely but not entirely on systems seen elsewhere. Defensive management styles – like the stultified and increasingly unhealthy family models of the Spencer iron syndicates – had been supplanted here by greater openness and innovative ethos. There were always possibilities.

But how premeditated were these new ventures? How consciously mutual? One clue is in the persistence of hand skills in certain functions, a 'best possible' but imperfect solution. There was indeed conscious enabling, direction and planning. Yet the scene quickly found its own dynamism. It generated new sets of trade interactions, new skills and workshop hierarchies. Just by existing, technology fulfilled an enabling role: the small cupola furnace allowed organizational change and autonomy while reinforcing links, so became both a cause and effect of innovation.

Ultimately, here was where two streams met: ironworking with chemistry, mechanical engineering with mathematics. And here was Scenius.

---

50 See ch. 5, above.

8

Perspectives

Let me not cast in endless shade
What is so wonderfully made

Alfred, Lord Tennyson, 'The Two Voices' (1833–4)

Respecting achievement

Weighing individual achievement in a balanced way – taking care with credit and attribution while questioning 'heroic' hyperbole – has guided this research. So has the idea of community as an incubator of industrial innovation. Those two forces, the individual and the collaboration, did not work in opposition. Together they demonstrate how innovation was supported within the era's most inventive industries. From that emerges a deeper understanding of what innovation was, and how it occurred, a credible and realistic view inside industrialization.

At the start, innovation's complexity was underlined. In concluding, the aim should be to keep things simple, and so it will be, as far as that is possible with a complicated subject. The concluding thoughts relate most directly to situations in eighteenth-century Britain, though they might conceivably have wider relevance, to other centuries and places. Some points apply to modern innovation, to industrial regeneration, to developing economies. Issues such as securing supply chains, adapting to new energy sources, developing skills, migration, and the meanings of community are high on contemporary political agendas. The experience of British industrialization has modern-day consequence: it was foundational to what we have today, and its story has potential to help address current dilemmas.

The ambition has been to draw back the veil cast by Victorians over this subject; and to argue for historical approaches as the way towards authenticity in this essentially human story. Here, there is something to say about Adam Smith.

Adam Smith and pin-making

The opening section of *Wealth of Nations*, about the division of labour, famously uses the example of pin-making.[1] Those few pages continue to be a source of confusion about the world of work in his time and place. Adam Smith and the pin factory demand perspective; the story is a neat illustration of the perils of not taking care about contexts. Here is the point: Smith was an economic thinker, not a seeker of industrial information. He was positioned, by birth and intellect, within a European élite, within the circles of philosophers discussed above in chapter 4. The pin factory was not at all as it first appears.

Approached historically, Smith's account of pin-making, and what it represented in terms of contemporary British industrial practice – that is, before *Wealth of Nations* was published in 1776 – invites questions. How typical at that time was his pin-making workshop? Was division of labour something new? And – as Smith's work and reputation were subsequently turned to the service of politics – how much of the retelling has distorted his own message? The story of Smith's pin factory, it turns out, says more about his own situation among European intellectuals than it does of British industrial process in the 1760s and 1770s.

Perhaps the workshop Smith describes was one he visited in Birmingham, then a major production centre of pins, nails, and countless other precision metal parts?[2] But that was not so; in fact, his information came from France. What is more, the pin factory most probably never existed. Smith's account was taken from Diderot's *Encyclopédie*, published in the volume for 1763 under the title 'Cloutier d'épingles'. 'Cloutier' means 'nailer': could the phrase mean a nailer who made pins; or be a misprint for 'clouterie', so a pin nailery? Neither actually makes sense, and nowhere does it say 'pin factory'.[3]

The pin-making example (both in Diderot and Smith) was a model, used to illustrate a general principle. Kay has shown that the claimed daily production could not have been achieved, and nor could the fine division of labour that Smith describes. Quite remarkably, Smith's base was in Kirkcaldy, close to the famed and innovative Carron ironworks – which he never visited.[4] This is not problematic, if we recognize that direct investigation of industry was not part of his method. As an intellectual, Smith's compass led him to French *économistes* rather than British grassroots. Smith's business was to explain economic

---

1 Smith, *Wealth of Nations*, I, Book 1, ch. 1, 5–15.
2 Division of labour in such trades is described by Berg, *Age of Manufactures*, 267–9.
3 Diderot (ed.), *Encyclopédie ou Dictionnaire Raisonné*; *Dictionnaire Le Petit Robert*. For older usages of 'clou', see Furetière's *Dictionnaire Universel*.
4 Kay and King, *Radical Uncertainty*, ch. 14, 'Telling Stories through Models', esp. 249, 253; Kay, 'Adam Smith and the Pin Factory'. With thanks to Matthew Ford for assistance.

theory, not to describe 'what was'. He deployed the story of a theoretical pin works, in a way acceptable in economics but baffling to historians.[5]

What Angerstein had witnessed in Birmingham – more than twenty years before Smith published *Wealth of Nations* – was indeed a division of labour, though bearing little resemblance to Smith's model. Angerstein illuminated the growth of skilled specialism, its purpose to enhance precision as much as increase output.[6] The dispersed factory arrangements, later enabling machine-making to grow from humble yet skilled origins, were an extension of what Angerstein had observed.[7]

Smith's singular approach explains what – to a historian – appear to be slips: his confusion of dexterity with skill, and not distinguishing between levels of skill; treating pin-making as a new trade; not recognizing the multiple options for organizing workshop-based production; missing the point that division of labour was already an interim and partial measure enabling 'best possible' production ahead of full mechanization. He correctly speculated that the drudgery of repetitive tasks stimulated innovation by 'common workmen', though not the importance of communality – the 'scene' – in technological development.[8]

But Smith was developing a thesis, not offering industrial guidance. Historians (in seeing elements of his work as a historical source) have misunderstood that. Meanwhile, the world in general has been distorting his legacy. Smith symbolizes how a message can be confused or lost in the noise of subsequent commentary. As with other characters discussed in previous chapters here – some very eminent, others far less so – his reality has been lost, through over- and under-interpretation. Smith, and all the others, are reminders that the answers historians pursue rest in their own time. Historicize the thing.[9]

Wealth of Nations was an enquiry into the meaning of wealth. Later, the ghost of Adam Smith was deployed to popularize 'a form of laissez-faire liberalism of a negative variety that exalted individual thrift and self-help', in the words of Winch. This version, strongly echoing Smiles, was in full popular swing by the time that the *Wealth of Nations* centenary was celebrated in 1876.[10] Smith himself may have been an altogether more sympathetic character. He saw advantages, for instance, in artisan practice, and viewed putting-out systems as exploitative.[11]

---

5   Kay, 'Adam Smith and the Pin Factory'.
6   See pp. 32–3.
7   See pp. 184–92.
8   Smith, *Wealth of Nations*, I, Book 1, ch. 1, 5–15, esp. 11–12.
9   See the introduction to ch. 3, above.
10 D. Winch, 'Adam Smith (bap. 1723–d. 1790), moral philosopher and political economist', DNB, esp. section 'Reputation'.
11 Berg, *Age of Manufactures*, 73

Whatever Smith was, he was not a recorder of the industrial world around him, in the mould of Angerstein and other travellers. Is it of any real concern that the pin factory did not exist? To economists, it seems not. Smith was laying down fundamental economic principles. His work is best compared with Lewis's concurrent efforts to systematize chemical elements. Adam Smith's conjecture that the tedium of repetitive tasks would spur shop-floor workers to find mechanized solutions was correct. It was, though, already occurring, so not much of an imaginative stretch on his part. But even to the greatest theoretician or practitioner, little beyond the very immediate future, past the next move, could then be seen.

Reflections on innovation

On that larger question, the future shape of innovation as the eighteenth century might have imagined it, it is hard to see how contemporary predictions could have been anything better than educated guesswork. Even the near future was hazy. No one could have said quite what was to come.

The British iron trade's transition to mineral fuel illustrates this. Looking back from 1800, the transformation must have appeared breath-taking. Yet even a decade earlier, there was still no certainty that a blanket adoption of coal-burning technologies was coming. Right up until the 1790s, the trade's emphasis had been, in the words of Evans and Rydén, 'stretching the boundaries of the "organic" economy'.[12]

Nothing was settled, until it was; and even after that, events might take an unexpected turn. Because industrial innovation encompassed technology, processes and systems, power sources, products and materials, premises, and more, one new factor potentially generated considerable knock-on effects. No surprise, then, that change was disorderly. Clearly it was occurring, in many industries and across most regions. But with no comprehensive overview of how the industrial world was advancing, many decisions about innovation must have been made, if not quite instinctively, then relating wholly or mainly to direct personal experience. Inevitably that was limited, perhaps to a localized production chain. Evaluation of technologies was difficult, suggesting that industrial choices could be haphazard.

But not all who wander are lost, and the growing tide of extemporization delivered useful outcomes. Pro tem measures, even quite minor tweaks, might sustain new systems until longer-term answers were found. To a modern observer, an advance that appears a technological cul-de-sac was perhaps, in its day, the best possible solution to a gap in a production system. That was not

---

12 Evans and Rydén, *Baltic Iron*, 40.

futile. The younger John Cockshutt, writing to William Lewis in 1763, referred to a small adjustment to existing technology as a conveniency, the context suggesting that such developments were routine.[13]

Here was an Age of Contingency, supported here and there by tentative scaffolding. Frequent readjustment was a necessary stage towards settling new technologies. There were many routes (recalling Watt's comment on Arkwright, without apology for repetition) to 'the making of it usefull'.[14] In certain instances, this extended over decades, constant modifications cumulatively transforming the original concept. Crompton envisaged his mule only as an aid to domestic spinning, but others quickly spotted its wider potential. It was not Crompton's for long. His method of drafting was transformational once rollers became adjustable, and again when automatic operation was introduced.[15] This is an example of a long evolution while in widespread use, and beyond the originator's influence. Cort, Cartwright, and Watt himself, were among the famed creators whose conceptions were by no means the finished item.

Increasingly it fell to machine-makers to progress a concept, by solving some of the inherent problems in prototype technologies and systems. False starts could contain the germ of a good idea. Perhaps that would thrive in a different context, just as electric cars now demand better charging networks and novel types of batteries, but could yet be another cul-de-sac. New materials expanded possibilities. Machine tools revolutionized how machine-making organized itself, the industry becoming largely factory-based, a generation after itself enabling the move of textile manufacture into factories.

Here is a strange paradox, that the trade of machine-making germinated from handmade parts. Making machinery was not itself a fully mechanized process before the advent of a comprehensive range of machine tools after 1815. A kind of community network took some of the strain. This improvisation was very evident in Keighley, and not quite so much in the busy industrial scene of Leeds, but present and important nonetheless. 'Productive association', or 'flexible specialization', is the division of manual labour into a fine art, to achieve the best possible precision in tasks then unmechanizable. It was part of a sub-system of supportive technologies, odd extemporizations recruited to plug holes within a more advanced production chain.[16] Among these were the

---

13   BL, Egerton MS 1941, ff. 5.4, 2 June 1763. 'Conveniency' is an obsolete form of 'convenience': OED.
14   See above, chs 3 and 5.
15   Catling, 'Development of the Spinning Mule', esp. 37–8; *Spinning Mule*, esp. ch. 3. Thanks to John Styles for ideas.
16   The manufacture of card-clothing is one important example of imbalances in textile production (others are power-loom weaving and wool-combing). See Cookson, 'Mechanization of Yorkshire Card-Making', esp. 50–1.

cupola furnace, the long persistence of Newcomen engines in specific contexts, and the need to nurse Boulton & Watt's engines into life.[17]

One striking ad hoc arrangement – another paradox – was the surge in handloom weaving during the second quarter of the nineteenth century. Even the most technologically advanced branch of textiles, cotton, employed about 250,000 handlooms in 1835, contrasting with the 106,000 power looms then at work. The powered version had been available before 1810, and yet, 25 years on, was still outnumbered 5:2 by handlooms.[18] Not all of those quarter of a million handlooms could be hangovers from earlier times: this was a peak in handloom working, and a sizeable number of them, though seemingly archaic, must have been recently constructed. Baines and Ure, observing 1830s manufacturing from different political perspectives, wrote of the wretched conditions in which hundreds of thousands of handloom weavers – many new to the trade – subsisted.[19]

This is an example of deploying supposedly obsolete techniques to address a failed link in an otherwise new or restructured production chain. The deficit here was created through advances in spinning and preparatory processes, and handlooms were used to avoid what would otherwise have been a paralysing bottleneck in weaving. This solution, obviously imperfect, continued through the time of the power loom's 'rather leisurely triumph' between 1815 and the 1850s.[20]

But situations were rarely simple, and neither was this. The textile industry – in its different branches and great variety of workplaces – was far from standardized. Many businesses embraced change, while others held out against it. According to circumstances, hesitancy could be entirely rational. Decisions came down to economics, but that encompassed quality concerns, logistics, and more.

Financial and productivity considerations are well described by Lyons, especially on the profitability of pre-1835 power looms.[21] Choices were not, as we have seen, made solely on economic grounds. The first power looms, from the 1810s, were quickly adopted for lower-quality cotton cloths, while makers of finer grades continued to favour handlooms for quality and economy. This was a

17 See above, pp. 58, 91.
18 Baines, *History of the Cotton Manufacture*, 382–3; Cookson, *Age of Machinery*, 18–20, and *passim*; 290–1 for estimates of machines at work after 1835. Farnie and Jeremy (ed.), *Fibre that Changed the World*, 25, confirm the figure of 106,000 cotton power looms in 1835. A power loom would of course be far more productive than any handloom, so the 5:2 ratio does not apply to the quantity of product.
19 Baines, *History of the Cotton Manufacture*, 493–502; Ure, *Philosophy of Manufactures*, 7–8, 338.
20 Lyons, 'Powerloom Profitability', esp. 392–3.
21 Lyons, 'Powerloom Profitability', is an excellent summary of his differences with von Tunzelmann, *Steam Power and British Industrialization* (esp. 195–203).

technological choice, with cotton manufacturers first to confront it. Afterwards, in the usual order of trickle-down, worsted and then woollen-cloth makers were presented with similar options. The self-acting looms available soon after 1840 finally made powered weaving standard across all textile branches.

Part of the logistical challenge was in creating capacity to produce sufficient power looms to meet demand, a task which the machine-making industry rose to address with its own machine-tool and factory revolution from the 1820s. But when did textile manufacturers accept that it was the right time to adopt and adapt, that a settled technology had arrived? Quite reasonably, not every business wanted to be first in line to try out something still emerging, incomplete. Others had a heavy weight of habit, or pulled back because in their own situation there was logic in sticking with older methods. The personal calculation may weigh against heavy investment in re-equipping factories with new-style machinery. Such rationalizations trumped the lure of latest innovations.[22]

The overlaps in forms of technology, the old running alongside the new, countered systemic imbalances. The mismatch took many forms. For instance, some power looms built in the 1830s were framed in wood, rather than using the cast-iron frames already widespread.[23] Such incongruities were not confined to textiles. As noted, the peak in water-power usage, as with handlooms, occurred some long time after apparently viable alternatives had presented themselves.

Managing transition threw up these disparities when resources were somehow constrained – perhaps by geography, production or technological bottlenecks, or shortage of skilled operatives. The economic calculations differed according to local or personal circumstances, whether in booming new sectors or longer-established industries.

But the constant, at times prolonged, attempts to fix these discrepancies continued. On the efficiency of waterwheels, some noteworthy examples illustrate how context and relevance were pivotal to meaningful advance. Desaguliers may have been correct in determining overshot wheels as most efficient, but there was no discernible follow-through. Smeaton, however, observing the growing shortage of watermill sites in his own region, proved the same by mathematical experiment and then, crucially, disseminated those findings. His theory was converted into an unusually precise and practical solution, a freely available site-specific template helpful to any developer of a water-driven site.[24] At Coalbrookdale, Abraham Darby II ran trials in his own works which demonstrated that supplementing water power with a steam pumping engine would enable year-round running of his blast furnaces. And so it continued. Displayed in Leeds in 1790 was a convergence of many ideas: some premises used rotary steam engines, others water power, while some large,

---

22 Hyde, *Technological Change*, 63–75.
23 An example, thought to be of the 1830s, is in Calderdale Industrial Museum.
24 See above, pp. 86, 135–6.

newly built cotton- and flax-spinning factories on the Aire were driven by waterwheels assisted by pumping engines. That proved a brief transitional measure, soon converted fully to steam. Later, thirty miles to the north in Nidderdale, far from the coalfields, the newly built Glasshouses Mill launched in 1812 as a fully water-powered flax-spinning factory. Furthermore, William Fairbairn updated Glasshouses in 1851 with a gigantic 120hp suspension wheel.[25] There were always local calculations to be made, as with the Somerset cloth trade before 1800, when persisting with water power had seemed a good option, and the coal canal arrived rather late.[26]

While hindsight is difficult to eliminate from such modern speculations, in certain conditions adoption lag was logical. The go-ahead Walker brothers powered their ironworks by water, the River Don a profuse and free supplier of power. Their first steam engine came only in 1782.[27] In Cornwall, where the mining sector had pioneered Newcomen engines from 1710, the last of that type was built in 1778. Many more than a hundred such engines had been installed in the county, consuming great quantities of coal brought in from other regions. As soon as Watt's more fuel-economic engine was available, the Cornish industry switched.[28] In contrast, the transformational iron-puddling process did not fully eradicate stamping and potting, which lived on in places, as it had value in certain circumstances – just as handlooms continued to have a place in specialist applications.[29] What appears as technological drag perhaps had its own sound reasoning. Innovation generally ran ahead of institutions – patents, regulatory bodies, infrastructure. Local economics were significant in favouring initiatives to suit their own conditions and technologies.

Elements within the status quo worked in favour of entrenching older technologies and systems. This was not sentiment: a technological decision represented a commitment (at least temporarily) which weighed against change. Once invested in technology (plant), or knowledge (a patent), the natural inclination was to bear with it. Prior investment was potentially an impediment if new and better options appeared. Patents themselves had negative impacts. The defence of patents at times blocked further advances, the kinds of upgrades and adjustments which improved performance. Cancelling Arkwright's patent in 1785 had unleashed a frenzy of activity in cotton-spinning and in spinning-frame innovation. A similar effect followed the expiry of Boulton & Watt's

---

25 See Giles and Goodall, *Yorkshire Textile Mills*, 125–36, 208.
26 See above, p. 17.
27 See above, p. 52.
28 John Kanefsky, paper on pre-1778 Newcomen engines in the UK (International Early Engines Conference, Elsecar, May 2017); and his database https://coalpitheath.org.uk/engines/index.
29 Hayman, 'Shropshire wrought-iron industry', 58–61, 84, 108–13; Hyde, *Technological Change*, 83–5.

monopoly on rotary steam engines and condensers in 1800. Their fierce protection of the patents had inhibited much-needed upgrading of what, by the late 1790s, was an increasingly tired technology. Matthew Murray and others tried to move matters along, adopting various ruses to circumvent Watt's monopoly and litigation, but not always successfully. Watt's restrictive policy played out in his own business, too. In dismissing initiatives – improvements in steam engines, and later a range of other innovations – by his loyal employee, the great technologist William Murdock, James Watt inhibited Boulton & Watt's technological options.[30]

There is a weight of evidence of collaborative working in many forms. Indeed, those legendary foreign 'spies' seeking British secrets were in many cases specialists engaging in cross-border co-operation. These agents sought to improve systems of supply or production, shared technical data, and discussed mutual concerns about external and political issues.[31] As for resistance to new technologies, another topic peppered with myth, the incidence has been shown as having been greatly exaggerated, and the purposes misinterpreted. Machine-breaking happened, but only exceptionally targeting a specific technology. Rather it was a stance in industrial disputes, one used especially against unpopular masters.[32] Thinking about the numbers of handloom weavers who were very recent recruits to the trade, tradition was not an issue, whereas pay and conditions were. As with child card-setters, no worker could be sorry to see the end of drudgery and exploitation.[33]

Dynamics of industrialization
And so to the controversy in economic history sparked by Robert Allen's contention that a high-wage economy – specifically, spinners' high pay – drove textile innovation during this period.[34] Recent discussion has considered the detail of industrial practices and relationships, connecting the debate into agency, changing social connections, and wider cultural and institutional contexts.[35] Humphries and Schneider, with others, mounted a robust challenge to Allen, using contemporary sources to reveal spinners' actual earnings. Evidence from, for example, the copious records of Oakes of Bury St Edmunds, prominent Suffolk clothiers, showed extensive spinning networks still at work in the closing decades of the eighteenth century, and wages set low by a regional

---

30 See pp. 126–7.
31 See ch. 2, above.
32 Cookson, *Age of Machinery*, 188–95.
33 Cookson, 'Mechanization of Yorkshire Card-Making'.
34 Allen, *British Industrial Revolution*; Ashworth, *Industrial Revolution*, 2–3; Hahn, *Technology in the Industrial Revolution*, 79.
35 See Berg's concise summary in *Age of Manufacture*, 180–2; also Kelly, Mokyr and Ó Gráda, 'Could Artisans Have Caused the Industrial Revolution?', 28.

cartel of manufacturers.[36] John Styles unpicked the economics behind the development of Hargreaves' spinning jenny, and found it to be rooted in quite local circumstances.[37]

Those approaches, searching within industrialization to understand specific contemporary situations, show the potential of archival research to rebut generalized claims which do not ring true. Those who challenge Allen are also supported by contexts, by understanding how innovation unfolded in real time. In the Suffolk instance, master clothiers had remained over-reliant on old systems and a low-skilled, badly paid workforce. They must have feared for their trade, aware as they were of the well-advanced technologies in northern England, but were powerless to react meaningfully.

Low wages impeded the introduction of textile technologies in East Anglia, because masters were invested in a putting-out system which served current interests. As Smail says, for some time they were secure, and stuck. This does not, though, mean that high wages drove innovation: in northern England, where low wages and poor quality were features of putting-out, that system was collapsing under its own weight. That proved a major impetus to press forward in search of technological solutions, a phenomenon already well underway by the 1770s. In contrast, the eastern counties showed little will or capacity to engage. The race was not lost, for it had never really started.[38]

Collectively, local experiences illustrate that there was no one unique pathway towards industrialization. The contrast between the Lit. & Phil. societies of Manchester and Leeds, northern England's most progressive and successful industrial towns, shows this. Their staple industries were similar, but the needs and opportunities of Manchester's cotton trade demanded something different, a one-off, which was duly created.[39] The ways of different localities are striking in their diversity, in how the present overwrote their past practices. The measure here is not simply one of economic success, nor purely in terms of positive or poorer outcomes. Shrewsbury and its region are shown here as the setting for feats of innovation and inspiration, a very significant source of manual skills, and well connected into channels of knowledge. This was also a place hampered as well as blessed by its own geography. None of those features is surprising, though, for many places displayed something similar. Those local experiences are always interesting, in themselves and assembled into the cumulative picture.

It has been shown conclusively that innovation was widespread, across many regions of Britain; and so too was awareness of the fact. This fed into a strong and rising confidence about what could be achieved.

36 Humphries and Schneider, 'Losing the Thread', summarizes the 'high wage' debate.
37 Styles, 'Rise and Fall of the Spinning Jenny', esp. 226–7.
38 See p. 17, above, for the fate of East Anglia; also Cookson, 'City in Search of Yarn'; for the failure of Yorkshire's putting-out networks, Cookson, *Age of Machinery*, 25–8.
39 See ch. 5, 'Philosophical societies'.

The scene
All history – of industry, engineering, science, politics, anything – is rooted in the dynamics of people's relationships. Change was disorderly because humans drove it, acting as humans do, with varying levels of imagination and initiative, and a range of motivations. Their endeavours were shaped by their own situation, and by the resources they could access. Much of this rested on the communities to which they were attached. Community does not have to mean locality, though it can be localized. Nor is 'community' the same as 'grassroots'. It is best thought of as people coming together and embarking upon a quest for solutions, perhaps to an industrial problem of more than individual concern, or maybe to commission infrastructure such as a river navigation. How did people in combination accomplish such schemes? It was a period of possibility, but in many of these ventures resources were limited, even by the standards of the time.[40]

Certainly there were exceptional talents at work in the middle of this complex picture. Awarding due credit, though, needs a pause, a reality check. No one starts from a blank sheet. Everyone benefits to some degree from what society and community offer, in knowledge, skills, and more tangible contributions. And even the great breakthroughs are not sacrosanct: all are improved, or at some point superseded by something better. The convenient shorthand of calling an innovation 'Arkwright's' or 'Cort's' distracts from the periods before and after, of development and enhancement.

Rather, there is the 'scene', which was actually many, sharply contrasting, scenes. Some of the most noteworthy influenced the evolving collision of science and technology, Dalton leading on chemical theory while his Manchester associates developed practical solutions to bleach cotton and light factories by gas.[41] There was the transfer of skills and industrial structure from south Yorkshire into Leeds's new industrial districts.[42] There was Scotland, with its distinct approach: university scientists involved early in industry; unusual public intervention to modernize industrial processes; and how Scottish thinking fed into the workplaces of northern England, through the education of nonconformists to whom English universities were closed.[43]

As older 'scenes' faded, replacements coalesced. The energy of a bright young man with autonomy, ideas, and cash – even limited amounts – must not be underestimated. From the 1740s, Samuel Walker and his brothers found a new path into how to do business, building from the Wortley nail-making connection. Older combines fragmented, making space for new forms of ironworking in south and west Yorkshire. The ossifying old-style Leeds

---

40 For 'community', see ch. 7, 'Innovating communities'.
41 See ch. 5, 'Philosophical societies'.
42 See ch. 6 'Networks and migration'.
43 See ch. 7, 'A galaxy of Scotsmen'.

merchant community was enlivened by the dynamic Elams, knowledgeable about North American exporting, willing to produce new cloth designs to satisfy demand, and ready to take a calculated risk. These examples formed the basis of new and significant settings.

Nor should it be forgotten just how many influential innovations were developed by artisans within a shop-floor scene. This was explored at length regarding the Wortley district, and its forgemen's (and others') long-term impact on mechanical engineering, including in how premises were arranged.[44] The owner-manager model brought former artisans – and former ironworks managers turned owners – into close constructive relationship with agents, merchants, customers, or others encountered through business or family connection, of the class characterized as 'intermediate'.[45] The benefits included direct investment into the business, and technical or commercial guidance. It was increasingly likely as the century moved along that new owners of ironworks engaged hands-on in innovating (Darby earlier, then Cockshutt, Crawshay, Dawson). Those who were less personally inventive attended to other details, financial and technical (Samuel Walker and his successors). In the start-up textile-engineering shops of this early period, the owner-manager likewise had a direct hand in almost everything that happened.

Class continued to have consequence, and the social gap was instinctively acknowledged. Yet in key trades which stepped up from the mid-century to meet industries' new demands, whether reconfiguring (ironmaking) or something new (machine-making), there is a sense of skill being recognized, talent acknowledged. Shifts in the boundaries of social acceptance are not measurable, but would Newcomen's feat have been so contested had he achieved something equivalent a generation or so later? Probably not, especially if it had happened in an industrial district. The significance of skill came to the fore with newly emerging technical specialisms. The smith identified a niche, closely focused his skills on specific components in great demand, built a business based on that, and, in becoming indispensable, commanded a degree of respect. That did not, though, deliver social parity, or not automatically.

Many significant innovators prominent during the second half of the eighteenth century had little appreciation of scientific theory, nor even much basic education. Remember the social torments of George Stephenson, because of his poor writing and rough speech.[46] Shop-floor workers and artisans often lacked even the basics of literacy and mathematics, working far below educational levels taken for granted in more élite circles. Without those basics, the idea of going into business independently – maintaining accounts, recording orders and technical information – was daunting, though some managed it.

---

44 See p. 190.
45 For 'intermediate', see the introduction to ch. 4.
46 See p. 102.

Was possession of knowledge an equalizing force? Perhaps for some individuals, but there is no indication that learning became less exclusive. But nor did privileged kinds of intelligence make much inroad into grassroots industry, at least not into the most significant innovative sectors. There were few obvious channels to convey it, and élite knowledge was not, for the most part, thought relevant to key industries. Potentially useful know-how may have been missed through failures of communication. It does seem, though, that even the groups with wealth and power remained frustrated by the limitations of their own useful knowledge. The historical reality of industrialization is that many of its leading lights were socially excluded from formal, packaged, education.

The histories which play on popular preoccupations with secrecy and spying miss their target: for protectionism and exclusivity, look to schools and highbrow institutions, not workshops. Infusions of fresh thinking into northern and midland England's industrial life came from Scotland, not from the Royal Society. If nonconformists featured large in that movement, it was not on account of their religion itself: it was that they had been obliged to go to Scotland for their university education. And Scottish artisans and technicians made many of their own notable contributions to English innovation.

The picture is very fragmented. There were substantial variations between localities in how sciences engaged with industry, all shifting considerably with time. Mathematics and mechanics, while comparatively well developed, were not accessible to most machine-makers. Smeaton had been unusual in applying an experimentally derived theory to very practical usage, but that example related to a power source, not to mechanical engineering. Chemistry, emerging from alchemy and still underdeveloped in the latter half of the eighteenth century, evidently learned more from ironmaking than it could deliver in return. Here, though, was a chasm between Scotland – where practical links between academic chemistry and industry were strong, building on the earlier and unique public policy initiative, the Board of Manufactures – and the remainder of Britain.[47]

That exceptional Scottish experience is one of the more interesting stories of innovation to have emerged here. The knowledge derived filtered into English industry largely through a handful of wealthy nonconformists obliged to seek their university education north of the border – among them William Henry, Thomas Percival, John Roebuck, Erasmus Darwin. Those men took away ideas of how science and industry might profit each other, and though few in number were disproportionately influential. The physicians among them received at least a grounding in chemistry, the experience feeding into significant networks forming in England. Most notably, the Manchester Lit. & Phil. played a significant role in developing the science of chemistry.

---

47   See ch. 7, 'A galaxy of Scotsmen'.

It is not to be assumed that all Scottish industrialists grasped science's potential to transform their business for the better. Very many did not, as David Mushet's frustrations testify.[48] In his time – from the 1790s – the value of science to technology was becoming more obvious in iron-making. Still, most new technologies of the period did not grow out of science. Examples touted as science-applied-in-industry may have been nothing of the sort. Even in 1800, William Reynolds, Joseph Dawson, and George Augustus Lee were atypical industrialists.

Are there generalizations from the experiences of the ninety significant figures? The most successful of the larger industrial ventures (Crowley and Wedgwood are examples) were concerned with efficient organization, whether or not that included new technologies. Rationalizing production systems went hand in hand with improving the transportation of goods. That could even be within one complex, but there must also be better external connection, locally, regionally, and with export markets. In the eighteenth century that largely meant waggon-ways for minerals, feeding into canals and river navigations if geography allowed. Surveyors and civil engineers appear in numbers here, in the early century preoccupied with water transport, and later developing the mathematics to enable construction of fireproof factories, iron bridges, and other complex structures. Of course, where the surveyors went, so did information and understanding, and that was a two-way process. It came down, always, to fresh appreciations and viewpoints. Cross Street Chapel can be seen as the religious wing of the Manchester Lit. & Phil., but wherever conversations occurred, among those who were interested, knowledge flowed. The risk-accepting Quakers who blazed a trail in North American trading, in iron goods as well as woollen cloth, were of similar approach, even if not of one mind about slavery and armament sales.

Technical communication was key, within or between workshops, or across industries. The evidence of how progress was made in such instances – think of puddling iron, generating gas, spinning cotton by machine – points to some key factors. Solutions rested on a mix of manual skill, industrial knowledge, vision and initiative, and dogged determination. The world from which innovation emerged was an imperfect one, served by unsatisfactory resources. In transcending all those difficulties, high in importance was an encouraging context. Here was the 'scene', and it embraced many possibilities. It was metalworkers collectively improving techniques, textile engineers refining products in consultation with customers, the new style of ironmaster reorganizing works and tasks – and others of the many examples discussed. Managers with autonomy could be especially influential, in assembling combinations of talented individuals from different backgrounds. Mixing to share

---

48 See pp. 183–4.

technical knowledge acted as a leveller. Inventive cultures sparked their own dynamism and positive feedback loop.

Where in all of this is the individual innovator? There have been many examples of one-trick ponies, and of serial innovators. There were twosomes, the likes of Wyatt and Paul, combining a technologist with a closely focused investor. There were men like Arkwright and Crawshay who were determined to make a concept work, hiring and discarding technical assistants until a winning formula was found. Without doubt, some individuals displayed exceptional talents, though circumspection is needed here. Visionary innovators could be deluded, by early success, into believing that they could do anything, and, in becoming unhinged from reality, stumbled badly. Blue-sky thinking has no bearings. Richard Roberts was not the only great mechanical engineer to bring himself down, eventually ruined by over-extending into designing and building steamships, because he could not stop. That is at the far extreme of the spectrum from those businesses too cautious to risk new machines and systems.[49]

Belonging to a 'scene' had the advantage of tapping into a group which was qualified to critically evaluate proposals within a context, local or regional, or some specialist activity. Another benefit was, when solutions to an industrial problem proved elusive, being able to respect others' knowledge and insight, with a prospect – thinking of William Lewis with John Cockshutt – of working something out together. The relationships were enabling. The scene gave a grounding to innovation, in validating a proposal as technically and financially viable – or otherwise. It is no coincidence that the ninety or so individuals in the appendix collectively display exceptional range in their interests and achievement. More than that, their multiple and overlapping personal links are demonstrable. Here are trails of real and productive connection, these apparently of mutual and wide-reaching benefit. It is a substantiation of the value of a 'scene'. The strands (based on descriptive accounts in earlier chapters) show substantial evidence of collective achievement.

Still, this is only part of a story, one to which many other histories could be added. Inevitably, ideas spin off from the fringes of this research. William Lewis merits further attention, especially in light of the rediscovered – rather, reidentified – sources in Cardiff and Stoke.[50] So do Joseph Dawson, the unusually scientific industrialist, and the origins of his Low Moor Company. New business histories of the early iron trade would be welcome. Substantial collections including the Spencer Stanhope Muniments (SpSt) in Barnsley, and

---

49 Ideas here from BBC Radio 4 series 18 of 'The Digital Human', on the Messiah, in the context of today's tech industry: https://www.bbc.co.uk/programmes/mooobcmz. See R.L. Hills, 'Richard Roberts', in Cantrell and Cookson (ed.), *Henry Maudslay*, ch. 4.
50 Cardiff Lib. Manuscript Collection, MS 3.250; World of Wedgwood, Collect Chem. Experiments 39-28405 (1771); 39-28406 (1768).

the Wharncliffe Muniments (WWM) and Wheat Collection (WC) in Sheffield, have more to offer on the early-modern iron trade in south Yorkshire but also far beyond, from the English Civil War and through the period of transition to mineral fuels, when the economy and management of coppice (charcoal) woodlands was a major preoccupation.[51]

Very recently, Berg and Hudson's comprehensive overview of commercial life and the global economy has established the powerful influence of slavery and exploitation upon British industrial transformation. Beyond doubt, the labour of enslaved people, the bloated profits of slave trading and plantations, and all that went with that, contributed significantly to this process, in many guises. Berg and Hudson argue that slavery's role has been 'generally underestimated by historians' and call for 'a rounded examination in the light of accumulated research'.[52] This sets a context of wider colonial history, and the experiences of merchants engaged in foreign trade, the likes of Broadbent and Elam and others here, are key to answering those questions.

So this is far from an ending.

---

51 See also YAHS Industrial History Online: https://www.industrialhistoryonline.co.uk/yiho/index.php.
52 Berg and Hudson, *Slavery, Capitalism and the Industrial Revolution*, 7.

Appendix

Webs of connection

For some overall analysis of this list, see chapter 7, 'The web of connection'.

Angerstein, Reinhold Rücker (1718–60)
Swedish ironmaster, public servant, traveller, university-educated.
Links: Swedish iron trade/metallurgy; ironworks across Europe, including Darby, Crowley, Wortley, Pontypool.

Arkwright, Richard (1732–92)
Textile innovator noted for developing water-frame, a key technology in cotton-spinning. North Lancs/east Midlands.
Links: Strutt, Atherton, Darwin, Wedgwood, Boulton & Watt, Wrigley, Jessop.
DNB (J.J. Mason)

Atherton, Peter (1741–99)
Instrument maker in Prescot; then innovator in textiles, cotton-spinner, factory-builder, in Lancs, Flintshire, Liverpool. Helped Arkwright with water-frame, early partner of Philips & Lee, and at Chipping.
Links: Carr/Chipping, Arkwright, Lee/Philips, Boulton & Watt, Albion Mill.

Bage, Charles (1751–1822)
Surveyor turned pioneering structural engineer, built first fireproof factory, theorized on strength of iron frames. Derby/Shrewsbury.
Links: Telford, Marshall, Benyon, Strutt, Lee, Darwin, Hazledine. Former Quaker.
DNB (A. McConnell); BDCE (T. Swailes)

Banks, John (1740–1805)
Philosopher, lecturer, writer. Kendal/Manchester. Dissenter, attended Rotheram's Academy in Kendal. Interested in strength of cast-iron beams.
Links: Rotheram, Shelf ironworks (Halifax/ Bradford border), Percival, Wilkinson, Dalton.

Barnes, Thomas (1747–1810)
Dissenting minister, Warrington/Manchester. Co-founder of Manchester Lit. & Phil. and academies, social reformer.
Links: Percival, the Henrys, Dalton, and influential circles within his congregation at Cross Street Chapel.
DNB (D.L. Wykes)

Benyon, Thomas (1762–1833) and Benjamin (1763–1834)
Woollens-to-flax merchants and manufacturers, and bankers. Shrewsbury. Unitarians, BB a political radical. Partners of Marshall and sponsored Bage's Ditherington Mill.
Links: Marshall, Bage, Murray, Hazledine, Boulton & Watt.

Bertram, Wilhelm (d. c. 1740)
Steelmaker, from Remscheid, one of several Germans settling near Shotley Bridge, Derwent valley, c. 1687. Bertram established shear steel and double shear steel, for the finest edge tools, on Tyneside c. 1730.
Links: Hayford, Crowley; German steel technologies.

Black, Joseph (1728–99)
Academic chemist famed for theories of heat and many other discoveries, at Glasgow and Edinburgh universities; studied and consulted on industrial processes. Elected FRSE 1783.
Links: Watt, Roebuck, Robison, Reynolds (student); David Hume, Adam Smith; engaged with Dundonald's tar works, Carron works, Cort process and other industry.
DNB (R.G.W. Anderson)

Boulton, Matthew (1728–1809)
Hardware manufacturer and merchant in Birmingham. Interests in science. Famously Watt's partner and financier in steam-engine manufacture, and founding member of Lunar Society.
Links: Roebuck, Watt, Wilkinson, Darwin, Owen, Wedgwood, Reynolds, Priestley; Benjamin Franklin; Lunar Society.
DNB (J. Tann); also 'Lunar Society of Birmingham' (J. Uglow)

Brandling family (active c. 1670–19th c.)
Landowners with extensive coal-mining interests in N.E. England, and then Leeds.
Links: Murray, Burdon, Walker, Gothard, Peter Fairbairn; Brandling agents introduced N.E.-style railroads and other colliery technologies to west Yorks, notably Middleton railway.

Broadbent, Joseph (d. 1761)
Wortley nail merchant, then Sheffield merchant and banker, exported cutlery to Virginia. As partner of Cockshutt, helped finance his start at Wortley Forge.
Links: Cockshutt; active and well known in south Yorkshire, esp. in Wortley nail-trade; Quaker with family links to London and Scarborough traders/North American exports.

Burdon, Rowland (1756–1838)
Wealthy politician, banker, industrialist, sponsor of transport improvements in Co. Durham, including pioneering Wearmouth iron bridge at Sunderland, 1793–5, cast by Walkers of Rotherham. Married into the Brandlings.
Links: Brandling, Sir John Soane, Walker, Thomas Paine; prominent Freemason; Newcastle Lit. & Phil.
DNB (G. Cookson); BDCE (R.W. Rennison and J.G. James)

Carr, William (c. 1750–1834)
Early textile machine-maker in Keighley, background in Arkwright technology in Scorton, north Lancs. Built and managed Kirk Mill, Chipping, 1785.
Links: Hattersley, Smith; Keighley/Leeds engineering hubs; Catholic; Freemason.

Cockshutt, John I (1711–74); John II (1740–98); James (1742–1819)
Innovative ironmasters with scientific interests; James also civil engineer and FRS.
Links: The Wortley scene, Wilson/Spencer relatives, Thomas, Walkers, Ayton Forge (Broadbent/Bland) and North America; Lewis, Angerstein, Smeaton (Rennie), and presumably Jessop; Society of Arts; South Wales (Crawshay, Cort, Boulton & Watt, Pontypool).
DNB (G. Cookson)

Cort, Henry (1741?–1800)
Naval agent, then ironmaster, initial developer of iron-puddling process.
Links: Crawshay, Gascoigne, Carron, Cockshutt.
DNB (C. Evans)

Cranage, George (b. 1701); Thomas (1711–80)
Contributed to development of stamping and potting process at Coalbrookdale; later worked at Carron.
Links: Darby, Reynolds; at Carron, perhaps knew Roebuck. Quaker.

Crawshay, Richard (1739–1810)
Yorkshire-born, London-based, iron merchant turned ironmaster; to make Cort's puddling technique viable, took over and made Cyfarthfa by far Britain's largest ironworks.

Links: Cockshutt, Cort; presumed relative of John Crawshaw, Walker partner.
DNB (C. Evans)

Crompton, Samuel (1753–1827)
Hand-weaver in Bolton, whose concept of mule-spinning transformed the cotton industry.
Links: Made contacts only after his technology was already widely shared and he campaigned for compensation, supported notably by Lee and Kennedy.
DNB (D.A. Farnie)

Crowley, Ambrose (1658–1713)
Ironmonger, later ironmaster, innovator in industrial organization. Stourbridge/London, launched nailing and steel ventures in Sunderland (1680s), then Winlaton and the Derwent valley (Tyneside, from 1691), to supply navy.
Links: Yarranton (tin-plate), Dudley, Hanbury of Pontypool; presumably Hayford; Lloyd bankers of Birmingham; acquainted with Hooke; daughter married into Cottons (Spencer associates).
DNB (J.M. Price)

Dalton, John (1766–1844)
Internationally reputed experimental chemist (atomic theory and periodic table); also teacher, industrial consultant (esp. bleaching), hub of the Manchester Lit. & Phil. FRS, lectured at Royal Institution, a founder of BAAS, many honours including the French *Académie des Sciences*.
Links: The Henrys, Percival, Lee, James Watt jun., Barnes, Rupp; many Quaker friends in Cumberland.
DNB (F. Greenaway)

Darby, Abraham I (1678–1717); Abraham II (1711–63); Abraham III (1750–89)
Coalbrookdale ironmasters, renowned for castings, vertical integration of mining and ironworking, transport initiatives.
Links: Quakers, especially the Goldneys and Reynolds of Bristol; Ford; Wilkinson; Cranages; perhaps Dudley. Linked by marriage and friendship to industrial Quaker families in Bristol, Sunderland, and Liverpool.
DNB: I (N. Cox); II & III (B. Trinder)

Darwin, Erasmus (1731–1802)
Physician with wide-ranging interests, including scientific experimentation, literature, and social questions. Liberal views. Universities of Cambridge and Edinburgh. Networked across the Midlands and beyond. FRS.
Links: Strutts, Arkwright, Boulton, Watt, Warltire, Priestley, Bages and more. Lunar Society, Derby Philos. Soc. founder, Manchester Lit. & Phil., Freemason.
DNB (M. McNeil)

Dawson, Joseph (1740–1813)
Presbyterian, then Unitarian, minister, turned ironmaster at Low Moor, Bradford, serious interests in science (especially mineralogy). From Leeds, educated at Daventry Academy and briefly the University of Glasgow (with charitable support).
Links: Glasgow science; Mushet; Mill Hill Chapel, Leeds merchants (large bequest from Christopher Holdsworth; daughter married a Busk).

Desaguliers, John Theophilus (1683–1744)
French-born, London-based, widely travelled, Oxford- and Cambridge-educated natural philosopher, demonstrator, and lecturer. Isaac Newton's experimental assistant. Wide interests, published on theoretical and practical matters. FRS.
Links: Newton; Stephen Hales; Sir Hans Sloane nominated him for FRS; many aristocratic and royal patrons. Huguenot turned Anglican minister, prominent Freemason with Scottish/continental connections.
DNB (P. Fara)

Drabble, William (b. 1769) and Joseph (b. 1773)
Early established, short-lived machine-makers in Leeds, most notably central to a kinship network from Wortley which fed into Leeds and Keighley engineering.
Links: Jubb, Hattersley, Wordsworth, close relatives by marriage; Salt, possibly.

Dudley, Dud (1600?–1684)
Oxford-educated experimental ironmaster who trialled smelting with coal. Author of *Metallum Martis* (1665).
Links: Crowley, possibly Yarranton; perhaps related to Abraham Darby I.
DNB (P.W. King)

Elam, Gervase (1681–1771); sons John (d. 1789) and Emmanuel (d. 1796)
Quaker woollen clothiers, tobacco wholesalers, then cloth merchants, initiated export trade in Yorkshire cloth to North America and shook up Leeds merchanting.
Links: With Quaker businesses in Philadelphia and elsewhere.
DNB (Gervase, by N.C. Neill)

Fairbairn, William (1789–1874)
Born Kelso, trained as a colliery millwright on Tyneside, a renowned, wide-ranging innovative engineer in London and (mainly) Manchester. FRS.
Links: Very well connected. Royal Society fellowship supported by (among others) Rennie, Cubitt, Joshua Field, Babbage, Ure; Member of Society of Arts; Inst. Mech. Eng.; Manchester Lit. & Phil.; BAAS.
DNB (J. Burnley rev. R. Brown)

APPENDIX

Ford, Richard (1689–1745)
From Stourbridge, clerk at Coalbrookdale and interim managing partner after his father-in-law Abraham Darby I died. Successfully expanded into new markets and products.
Links: Quaker, married into Darbys; Goldney, Thomas Newcomen.

Garbett, Samuel (1717–1803)
Birmingham merchant, starting out in precious metals and manufactured goods, then chemicals (at Prestonpans) and iron (co-founder of the Carron works).
Links: Scotland; Roebuck, Cadell, Gascoigne (his son-in-law); Boulton, Watt, Wedgwood, Bage, Darwin; Wyatt and Paul; shares in Albion Mill. High-level political contacts. Knew the Lunar men, though unclear whether he attended. Interests in social issues in Birmingham (municipal affairs, societies, charities).
DNB (R.H. Campbell)

Gascoigne, Charles (c. 1738–1806)
After marrying Mary Garbett, moved to Carron as managing director in 1763. Noted for creating a new kind of plant, overcoming technical issues, building links to Russia while supplying armaments. After bankruptcy, took a group of Scottish key workers to Russia in 1786 and did not return.
Links: Roebuck, Cadell; Garbett and his circle; Smeaton; probably Murray (re engine exports to Russia c. 1800).
DNB (E.H. Robinson); BDCE (M. Chrimes)

Goldney, Thomas III (1696–1768)
Merchant concerned in diverse industrial and mining concerns in the Severn Valley (esp. Coalbrookdale), N. Wales and S.W. England, alongside his family's Bristol-based commercial interests.
Links: Darby, Wilkinson, Ford, Reynolds; Bristol Quaker circles.
DNB (K. Morgan)

Gothard, Timothy (1723–1805)
Ironworks millwright turned ironmaster, progressing from Wortley to Sheffield, Rotherham (within Walkers' orbit), then Leeds in 1770, where he established the town's first cupola furnace at Hunslet.
Links: Wortley group, particularly Jubb; Salt, Walkers, Middleton Colliery.

Gott, Benjamin (1762–1840)
From a family of surveyors, famed for Bean Ing Mill, an integrated woollen factory (though largely using hand processes) built to support his merchant house.

Links: Leeds commercial circles; friend of Watt, Rennie, Lee. Interested in art; first president, Leeds Phil. & Lit. (1818); a founder of Leeds Mechanics' Institute (1824).
DNB (R.G. Wilson)

Hattersley, Richard (1761–1829)
Whitesmith from nail-making background in Wortley. Moved to Sheffield, Kirkstall (making large screws for machines), then Keighley. There built a hub of component-makers supporting textile machine-making.
Links: Wortley family connections including Drabbles, Jubb, Wordsworth; same links later in Kirkstall/Leeds, and in Keighley scene; Swedenborgian, Freemason.

Hayford, Denis [or Heyford] (c. 1635–1733)
Pioneered steel-making on Tyneside through the 'Company in the North'. Perhaps from Silkstone (Wortley). Multiple interests in forges and furnaces, in Yorkshire, across the Midlands, Cheshire, and near the Tyne at Shotley and Blackhall.
Links: Part of Spencer syndicates and their associates; Bertram; presumably Crowley; Cotton, Simpson, Fell, all Spencer associates and his co-partners in Company in the North.
DNB (B.G. Awty)

Hazledine, William (1763–1840)
Shrewsbury millwright and ironfounder, produced innovative castings for Ditherington, the first iron-framed building, and became national leader in building iron bridges. Knew Telford as a Freemason, and contributed significantly to all Telford's great works, canals, docks, bridges, etc., nationwide.
Links: Telford, Bage, Jessop, connected into Reynolds network. Freemason.
DNB (S. Hughes); BDCE (Powell)

Henry, Thomas (1734–1816)
Apothecary, scientist, educationalist. Practical chemist into industrial applications, including chlorine for cotton bleaching. Co-founded schools including Manchester Academy, which brought Dalton to town and where Henry himself taught evening classes.
Links: Boulton, Watt, Watt jun., Priestley, Percival, Dalton, Lee etc. Widely networked through Unitarians/Cross Street Chapel; a founder and early secretary of Manchester Lit. & Phil. Radical politics. FRS 1775 (sponsors included Priestley and Percival).
DNB (F. Greenaway)

APPENDIX

Henry, William (1774–1836)
Physician, chemist, experimented on properties of gases. Involved Lee in gaslighting trials. Worked closely with his father Thomas, Percival and Dalton. Attended Manchester Academy and Univ. of Edinburgh. Concerned with disease and state of towns.
Links: Percival, Dalton, friend of Lee; Boulton and the Watts. Unitarian/Cross Street Chapel; very active in Manchester Lit. & Phil; supported BAAS. FRS.
DNB (F. Greenaway)

Hodgkinson, Eaton (1789–1861)
Structural engineer and material scientist. Tutored by Dalton. Experimented for Fairbairn on strength of cast-iron beams and of bridges. First prof. of mechanical engineering at UCL, from 1847.
Links: Dalton, Fairbairn, Robert Stephenson; Cross Street Chapel; very active in Manchester Lit. & Phil.; a director of Manchester Mechanics' Institute; BAAS; Institution of Civil Engineers. FRS 1841 (sponsors including Brunel, Joshua Field, George Rennie).
DNB (D. Cardwell); BDCE, vol. II (J. Sutherland)

Huntsman, Benjamin (1704–63)
Clockmaker, Quaker perhaps of German descent, from 1740s first to cast crucible steel into ingots, and of higher quality than the cementation steel then in use. Then established this process in works in Sheffield c. 1750. The only means of producing steel ingots before Bessemer (1856).
Links: Huntsman did not patent, but his works was small and secretive. Salt may have learned the process there, or with a local emulator.
DNB (D. Hey)

Jessop, William (1746–1814)
Canal, dock, railroad and harbour engineer with wider industrial interests. His father worked with Smeaton on the Eddystone Lighthouse, 1756–9, and Smeaton employed the son as pupil, then assistant, based in Leeds from 1760. Jessop ran his own business from 1772, at Ferrybridge and Newark.
Links: Smeaton; John and William Gott, surveyors (Benjamin Gott's father and half-brother); Telford, Rennie, Benjamin Outram; commissioned by Arkwright, presumably knew James Cockshutt; secretary of the Smeatonian Soc. of Civil Engineers.
DNB (R.A. Buchanan); BDCE (R.B. Schofield)

Jubb, John (1748–1808)
Ironworks millwright, strong Wortley (S. Yorkshire) connections, influential early migrant from there to Leeds (Churwell, c. 1778, central Leeds 1787–8),

establishing the town's first machine-making shop alongside millwrighting. Encouraged new wave of Wortley connections to settle in Leeds. Those he trained built a new generation of machine shops.
Links: Related to Drabbles, Hattersley, Wordsworth; close to Gothard; customer at Kirkstall; partner in the earliest local Arkwright-style cotton mill, with Wetherills, clothiers and fellow Congregationalists, at Churwell.

Lee, George Augustus (1761–1826)
Clerk, to manager, then partner in Manchester cotton-spinning concerns. Astute organizer and businessman. Introduced to Manchester Lit. & Phil. by business partners George and John Philips. Lee was not active in society affairs but connected with many other members. Constructed an early fireproof factory, then lit it with gas, working experimentally with Henry.
Links: Strutt, Bage, the Watts, Murdock, Gott, Crompton; partners Atherton, Philips; married into Ewart family; Manchester Lit. & Phil.; Unitarian.
DNB (J.J. Mason)

Lewis, William (1708–81)
Academic experimentalist, author and lecturer, supported by multilingual scientific assistant Alexander Chisholm. One known excursion from London, enquiring into the chemistry of iron-making at Wortley and Coalbrookdale.
Links: Cockshutt, Darby, Reynolds, Stephen Hales, Desaguliers, Charles Wood; international range of correspondents. FRS (1745).
DNB (F.G. Page)

M'Connel, James (1762–1831) & Kennedy, John (1769–1855)
Left Kirkcudbrightshire for machine-making apprenticeships with William Cannan, M'Connel's uncle, at Chowbent, Lancs. Set up in Manchester, pioneering steam-powered mule-spinning, but soon converted to large-scale cotton-spinning. Kennedy active in Manchester Lit. & Phil., with papers on technical and social issues; M'Connel supported Mechanics' Institute; Unitarian.
Links: Wrigley, Lee, Crompton; Chowbent (nail-making); William and Peter Fairbairn; Lancs/Glasgow cotton magnate Thomas Houldsworth (M'Connel's brother-in-law), and other connections in Glasgow.

Marshall, John (1765–1845)
Leeds linen merchant, employed Murray on mechanised flax-spinning trials. Loans from fellow worshippers at Mill Hill Chapel helped build his flax business. Later, he supported many educational initiatives, in Leeds and elsewhere.

Links: Murray, Benyon (Ditherington, Bage, etc.), Thomas Carlyle, Watt, Wrigley; his friends, Peter Garforth (Shipley cotton-spinner) and David Wood (Murray's partner). Unitarian, later Anglican. Founder member and first president of Leeds Phil. & Lit., and a founder of the Mechanics' Institution.
DNB (M.W. Beresford)

Murdock, William (1754–1839)
Ayrshire millwright, solver of engineering problems – in steam engines, machine tools, factory organization, and notably gaslighting, earning the RS's Rumford medal. Long career with Boulton & Watt, in Birmingham and Cornwall, and improver of their designs.
Links: Boulton, Watt, Lee (presumably also William Henry), Rennie, Nasmyth, Marshall, Joshua Field; James Boswell.
DNB (J.C. Griffiths); BDCE (R. Birse)

Murray, Matthew (1765–1826)
Newcastle-born smith, solved Marshall's flax-spinning problem c. 1789. In his own business, proved an innovative engineer: textile machinery, steam engines, locomotives and railroads, and early template for an integrated engineering factory with foundries. Supported Leeds public gasworks.
Links: Marshall, David Wood, Fenton, Blenkinsop (Brandling agent, 1811), Charles Gascoigne Maclea, Samuel Owen, Watt jun.; opened markets in Sweden and in Russia (which Murray may have visited, and where his son later worked).
DNB (G. Cookson); BDCE (M. Chrimes)

Mushet, David (1772–1847)
Major figure of his era in Scottish iron and steel, anticipating mid-century developments. Metallurgist, experimentalist, pioneer of technical writing, discoverer of blackband ironstone reserves in Airdrie. Patented direct steel-making in crucibles, and improvements to puddling. Innovation based in scientific investigation, findings published in the *Philosophical Magazine* and elsewhere.
Links: Dawson, Neilson; well known and connected into Scottish scientific circles.
DNB (I.J. Standing)

Newcomen, Thomas (bap. 1664–1729)
Dartmouth iron merchant, spent a decade with John Calley devising an atmospheric steam engine to drain mines in Cornwall, apparently without prior knowledge of Papin or Savery engines.
Links: West Midlands and Cornwall trade connections.
DNB and BDCE (J.S. Allen)

Owen, Samuel (1774–1854)
Carpenter and pattern-maker who became 'father of Scandinavian engineering'. From Shropshire, to the Soho works of Boulton & Watt, and then to Murray in Leeds. Installed engines in Sweden for Murray in 1804, moved there to work at first for a former Carron worker. Owen made the first steam engines produced in Sweden, and later, bridges, dredgers, and general castings.
Links: Watt, Murray and their employees.
BDCE (M. Chrimes)

Paul, Lewis (d. 1759)
Of French Huguenot ancestry, conceived roller-spinning, and the idea of selling licences for patented innovations. With his carpenter/mechanic John Wyatt, who overcame practical difficulties, Paul devised textile machines for spinning and preparatory processes. Roller-spinning, patented in 1738, was the basis of Arkwright's later innovations but never commercially successful for Paul.
Links: Assisted financially by some of Dr Johnson's Lichfield circle. Garbett invested but lost money.
DNB (R.B. Prosser rev. G.Cookson)

Percival, Thomas (1740–1804)
Educated at Warrington Academy (which he later relocated to Manchester), trained as a physician at the universities of Edinburgh and Leiden. Practised little, working instead on public health, epidemiology, sanitation and factory conditions, managing schools and hospitals. Founding vice-president of Manchester Lit. & Phil, then president until his death. Made FRS aged twenty-five.
Links: William Robertson, David Hume; Manchester Lit. & Phil., including Barnes, the Henrys, Dalton, Banks, Watt jun. Unitarian.
DNB (A. Nicholson, rev. J.V. Pickstone)

Priestley, Joseph (1733–1804)
Polymath, theologian, teacher, experimental chemist and writer. Elected FRS and made his name with *History of Electricity*. Moved to anti-Trinitarianism, accepted by his congregation at Mill Hill. In Leeds he did important scientific work, on electricity, optics, and then 'inquiries concerning air', which won the RS Copley medal in 1773. FRS.
Links: Warltire, Wilkinson, Hales. Widely connected through many moves around northern England and the Midlands, and finally Philadelphia. Notable: Lunar Society, and the intellectual group around Surgeon Hey in Leeds. Unitarian.
DNB (R.E. Schofield)

Rennie, John (1761–1821)
East Lothian millwright turned civil engineer, built bridges, canals, harbours, factories. Considered less original but wider-ranging than Telford. FRS, FRSE.
Links: Robison, Watt and Watt jun., Black, Ewart, Boulton, Andrew Meikle, Walkers, Fairbairn, Gott, Jessop, Murdock; Smeatonian Soc. of Civil Engineers.
DNB (A. Saint); BDCE (P.S.M. Cross-Rudkin)

Reynolds, Richard (1735–1816)
Interim managing partner at Coalbrookdale after his father-in-law, Abraham Darby II, died. Further developed company's transport and organizational arrangements. Philanthropist, anti-slavery. Father of William.
Links: Lewis, Cranage, Darby partner, Goldney, Wilkinson, Boulton, Watt, Wedgwood, Fry; fellow Quakers, reformers. Through father (Bristol iron merchant) he knew the S. Wales iron and tin-plate industry.
DNB (B. Trinder)

Reynolds, William (1758–1803)
Wide-ranging and much admired, noted for his vision of highly integrated operations at Coalport, using and recycling assets to best effect. Abiding interest in science and its industrial potential, ran tests/trials on processes and constructions in his works.
Links: Wilkinson, Black, Jessop, Telford, Watt, Crawshay, Banks, Lord Dundonald (significant partner), Trevithick, Hazledine. Retained Quaker connections and principles, though marriage to a first cousin excluded him from meetings.
DNB (B. Trinder); BDCE (J. Powell)

Robison, John (1739–1805)
'Engineering scientist', chemist, surveyor, structures, instruments (tested on expedition). Proposed Newcomen engine improvements, and encouraged Watt at Glasgow. Many scientific papers, including on electricity, optics, astronomy, and a planned text on practical mechanics for millwrights.
Links: Watt, Black, Telford, taught and mentored Rennie; friend of the Wedgwoods; met Darwin and other Lunar men. Northern Lighthouse Board; strong links with Russia, Quebec. RSE (founding secretary).
DNB (P. Wood); BDCE (M. Chrimes)

Roebuck, John jun. (bap. 1718–94)
Physician in Birmingham, said to have commercialized Scottish science with his and Garbett's industrial-scale production of sulphuric acid at Prestonpans. A founding partner of the Carron ironworks, held controlling share in Watt's

patent and may have introduced him to Boulton. Brilliant industrial chemist and metallurgist; short of business acumen and ended bankrupt.
Links: Garbett, Watt, Boulton, Gascoigne, Cadell family; educated Edinburgh and Leiden, close to major figures of Scottish Enlightenment; knew Walkers of Masbrough. Unitarian.
DNB (R.H. Campbell)

Rupp, Theophilus Lewis (d. 1805)
German cotton-spinner and chemist (unclear which came first). Built apparatus, produced 'recipes'. Published many papers on bleaching and chemicals in Manchester Lit. & Phil. proceedings; appraised Priestley's work.
Links: Manchester Lit. & Phil. circle including Dalton, Watt jun.; friend of William Henry.

Salt, Titus (1724/5?–1804)
Whitesmith who styled himself cast-steel maker when leaving Sheffield for Leeds in 1772. Joined Gothard as partner and main investor in Hunslet Foundry, which used cupola technology to remelt iron.
Links: Gothard, Jubb; perhaps Huntsman; may have had Drabble relatives.

Smeaton, John (1724–92)
Consulting engineer and scientific investigator of huge range and achievement. Much work related to water transport or water power. FRS.
Links: Jessop, Gascoigne, Priestley, Cockshutt, Rennie, etc., etc.
DNB (A.W. Skempton); BDCE (M. Chrimes)

Smith, William (1774–1850)
Mill mechanic, one of three early machine-makers in Keighley. Innovator in textile technology, built a large and successful business.
Links: Carr, Hattersley; Arkwright system.

Spencer partners (active c. 1630s–1760s)
Ironmasters and investors with interests across northern England and west Midlands. Dominant in Yorkshire charcoal iron, c. 1700–c. 1760. Wide kinship and agent network.
Links: Hayford, Stanhope, Dickin, Cotton, Wilson, Cockshutt, etc.
DNB (B.G. Awty)

Strutt, Jedediah (1726–97); Elizabeth (née Woollat) (1729–74)
Wheelwright, improved, patented and manufactured hosiery frames. Thoughtful and bookish Presbyterian/Unitarian. Elizabeth a persistent fundraiser for their early industrial enterprises, through nonconformist friends and networks in

London. Later, Strutt advised Arkwright, investing in his cotton-spinning enterprise as partner, though disapproving of Arkwright's business ethics.
Links: Arkwright; Thomas Cheek Hewes; Walkers of Masbrough. Unitarian circles.
DNB (J.J. Mason)

Strutt, William (1756–1830)
Son of Jedediah and Elizabeth, turned from cotton manufacture to become architect, engineer, and machine-maker. Worked on theory of cast-iron beams. FRS.
Links: Bage, Lee, Darwin (co-founders of Derby Philos. Soc.); Coleridge, Edgeworths, Bentham, Robert Owen, Marc Brunel, Samuel Oldknow. Unitarian.
DNB (J.J. Mason); BDCE (T. Swailes)

Taylor, Joseph (1777–1843)
Leeds-born, early apprentice to machine-making. With partner Wordsworth, built a large and successful engineering business.
Links: Presumably trained by Jubb in Leeds, then worked for Drabble brothers' who took over Jubb's first machine shop. He and Wordsworth married sisters, cousins of the Drabbles, connecting further into the Wortley diaspora in west Yorkshire.

Thomas, Callisthenes (active at Wortley by late 1730s and into 1750s)
Trusted adviser, agent, and man of business in the iron trade, commissioned by various Yorkshire ironmasters. Thought to have originated in south Wales (?Pontypool).
Links: Wilson, Cockshutts (Thomas later married sister of John I), Broadbent, Spencer, south Yorkshire iron-trade managing agents, e.g. Fell.

Walker, Jonathan (1710–78); Samuel (1715–82); Aaron (1718–77)
Of Grenoside and Masbrough/Rotherham. Ironmasters, suppliers of cast iron parts and steel, to a range of industries, and with interests in lead-mining.
Links: Close to the Wortley scene; a wide customer base. Burdon, Watt, Thomas Paine, Bage, Samuel Aydon; Gothard and Middleton, Leeds; Tyneside; three sons of Samuel II m. in 1780s daughters of Samuel Need, Nottingham hosier, partner of Jedediah Strutt.
DNB (D. Hey); BDCE (M. Chrimes)

Warltire, John (c. 1763–1810)
Swiss-born touring science lecturer and demonstrator, experimented on air, assistant to Priestley at Calne, afterwards tutor to children of Darwin and Wedgwood.

Links: Priestley, Watt, Boulton, Darwin, Wedgwood and Lunars; Derby Philos. Soc.; Samuel Johnson, Fry of Bristol.
DNB (H.S. Torrens)

Watt, James (1736–1819)
Famous for steam engine, but also surveyor, improver of instruments, later chemistry (bleaching, gaslight). Retained Glasgow connections, later developed significant new ones in Manchester. FRS.
Links: Extensive. Include Black, Robison, Roebuck, Boulton, Telford, Darwin, Wilkinson, Priestley, Rennie, Murdock and many more; Literary Soc. of Glasgow; Lunar Society.
DNB and BDCE (J. Tann)

Watt, James jun. (1769–1848)
Trained by Wilkinson in iron manufacture, drawing, carpentry; grounded in textiles in Manchester, where he found outlets for his interest in chemistry. Officer and experimentalist in the Lit. & Phil., with interests in bleaching and gasworks. FRS, supporters including Rennie and Humphry Davy.
Links: Manchester Lit. & Phil.; Soho connections and Lunar men; Murdock, Lee, Dalton, Percival, Henrys, Ewart.
DNB (E.H. Robinson)

Wedgwood, Josiah (1730–95)
Developed integrated pottery works, exporting by canal, organized business systems with cost-accounting. Scientific interests, especially in geology. FRS.
Links: Chisholm, Lewis, Warltire, Garbett, Robison, Arkwright, Reynolds, Watt, Priestley, Darwin. Lunar Society, Manchester Lit. & Phil. Unitarian and abolitionist.
DNB (R. Reilly)

Wilkinson, John (1728–1808)
From Workington, attended Rotheram's dissenting academy in Kendal, trained with a Liverpool merchant, took over Bersham ironworks from his father and bought other premises at Coalbrookdale. Innovative ironmaster, made cylinders for Soho, and was significant in developing iron technologies.
Links: Boulton, Watt, Watt jun.; Priestley (brother-in-law), Darby; brother of William Wilkinson; friend of Crawshay and William Reynolds. Society of Arts. Presbyterian turned Unitarian, political radical.
DNB (J.R. Harris); BDCE (M. Chrimes)

Wood, Charles (1702–74)
Ironmaster, with his brother John improved stamping and potting method of working wrought iron, and made other innovations. Interest in metallurgy,

introduced platinum from Jamaica and identified it as an element. Built first forge at Cyfarthfa in 1766, assembled an English workforce, and managed the works until his death.
Links: William Brownrigg FRS (who took the platinum sample to Lewis and the RS); active in west Midlands, Cumbria, Cyfarthfa.

Wood, David (1761–1820)
Tadcaster-born smith, basic education, perhaps employed by Marshall and met Murray there. Murray's first partner, a talented mechanical engineer who managed the works, freeing Murray to innovate in new fields, including steam engines and locomotives.
Links: Friend of Marshall; in Murray's circle.

Wordsworth, Joshua (1780–1846)
Carpenter from Wortley, married a cousin of the Drabbles, then moved to work for them in Leeds in 1803. Her sister afterwards married Taylor. After Drabble failed in 1812, Taylor and Wordsworth took their works and launched a highly successful textile-engineering partnership.
Links: Wortley group in Leeds and Keighley, including Jubb, Hattersley, and workers later brought in from south Yorkshire.

Wrigley, Joshua (fl. 1772–1810)
Manchester millwright, pump-maker, installer of modified Savery-type engines to supplement waterwheels, and of mill machinery. Reputed to have made more engines than Boulton & Watt at that time, though quality and efficiency were poor.
Links: Marshall, M'Connel & Kennedy used his engines, Smeaton tested them. Wrigley was a witness at Arkwright's patent trial in 1785.
BDCE (M.Chrimes)

Yarranton, Andrew (1619–84)
Ironmaster, river engineer, and surveyor, proto-political economist. Improved the Stour and other rivers; carried out trials producing tin-plate after witnessing the process in Saxony. Most noted for ideas on industrial improvements and advancing trade.
Links: Well connected in Worcestershire and London; Crowley (probably the first Ambrose Crowley), Dudley (though antagonistic); Robert Hooke.
DNB (E. Clarke rev P.W. King); BDCE (P.W. King)

Glossary of terms related to iron-making

Bar-iron	Wrought iron in the form of bars.
Blast furnace	From the late fifteenth century, blast furnaces smelted iron ore to make pig iron, which could if required be converted to malleable iron by forging. This, called the indirect process, used charcoal fuel in all stages, until the eighteenth century, when mineral fuel (coke) was introduced in blast furnaces, and coal was used in certain of the forging operations.
Bloomery	From prehistoric times and into the modern period, iron was made by a direct process. The bloomery, a single hearth fuelled by charcoal, turned iron ore into malleable iron.
Cast iron	Extremely strong in compression, though weak in tension, cast iron contains up to 4 per cent carbon, a far higher percentage than wrought iron. It was produced by melting iron to a liquid state, and setting in moulds.
Cementation steel	Also called German steel, the steel in general use before crucible steel.
Coking	Coke is made by 'cooking' coal into a purer form, making a fuel suitable for cast-iron production. It did not match the purity of charcoal, which endured in forging until the puddling method was introduced, when coke became the standard fuel.
Crucible steel	Benjamin Huntsman's breakthrough in the 1740s, known in his time as 'cast steel'. Huntsman adopted a technique used by brass-founders, melting steel in clay crucibles to achieve consistent quality.
Cupola furnace	A small remelting furnace used increasingly from the mid-eighteenth century, liberating iron-component manufacture to move away from the vicinity of large ironworks.
Forge	An open hearth or fireplace, with bellows. Here iron was melted and refined, hammered by smiths to remove impurities and converted to malleable iron.
Foundry	A place where iron was founded, or cast.
Pig iron	Iron tapped from a blast furnace and cast into 'pigs' or ingots, ready for remelting and processing into cast- or wrought-iron products.

Puddling and rolling	Initiated by Cort, the long-sought solution enabling coke-smelted pig iron to be converted to wrought iron using mineral fuels. It revolutionized the iron industry from the 1790s.
Reverberatory (air) furnace	A development of the seventeenth century, enabling pig iron to be refined in a chamber separate from the fuel. It improved iron quality, and was economic in reusing scrap iron in the mix. With these furnaces, iron goods could be produced away from a blast furnace.
Rolling mills	A process using rollers to flatten or shape iron or other metals. It was used to good effect in the puddling method.
Slitting mills	A kind of rolling mill used to prepare long thin rods for the nailing trade, first used *c.* 1600.
Smelting	The process of melting ore in order to extract the metal.
Stamping and potting	One of several attempts in the decades before puddling to refine pig iron using mineral fuel. Stamping and potting was developed at Coalbrookdale and patented in 1766.
Tin-plate	The tin-plate industry became significant after John Hanbury of Pontypool developed a method of coating rolled iron with tin, *c.* 1700. It concentrated in south Wales.
Wrought (or malleable) iron	For almost 2,000 years, this was the only form of iron available. Wrought iron, easily worked by hammering or rolling, is almost pure. It has a high tensile strength, is fibrous, and resists corrosion.

Further reading: Hayman, *Ironmaking*; Cossons, *Industrial Archaeology*, ch. 6.

Bibliography

Primary sources

Barnsley Archives and Local Studies Department [BALS]
Spencer Stanhope Muniments (SpSt)

Birmingham Library
MS 3219/4/80B Boulton & Watt papers

Borthwick Institute for Archives, University of York [BIA]
Probate of Anne Walker of Ecclesfield, July 1741, vol. 87 f. 433
Will of Christopher Holdsworth, Prerog. 1775

British Library [BL]
Egerton MS 1941 (1735–76)

Brotherton Library Special Collections, University of Leeds [Brotherton SC]
YAS/DD185/22; /31; YAS/DD233; YAS MS2022, John Brearley commonplace
    book (1772–3) https://www.yas.org.uk/Collections/Brearley-Memo-Book
MS 193/117A, Gott notebook of mill practice
MS 200, Marshall & Co., business archive

Cardiff Central Lib. Manuscript Collection
MS 3.250

Kirkstall Forge (Commercial Estates Group Ltd)
Plan of Kirkstall Forge by George Seed 1844

National Records of Scotland
NG1:   https://www.nrscotland.gov.uk/research/research-guides/research-
    guides-a-z/scottish-government-records-after-1707

Rotherham Archives [RA]
1263-Z Munford typescript
31-F/3/1, copy probate of Samuel Walker jun., 1792

328-B/1/1, 'Proposals' booklet, almost certainly the work of Thomas Walker, late 1783 or early 1784
Walker family ephemera, vols 1 and 2, including transcripts of inventory of Joseph Walker of Stubbing House, Ecclesfield, d. 21 Dec. 1729, and of his will, 21 Oct. 1729
Annotated pedigree of Walkers based on Foster, *Pedigrees of the County Families*

Science Museum Lib.
Goodrich mss, GE2

Sheffield City Archives [SCA]
Wharncliffe Muniments (WWM)
Wheat Collection (WC)

The National Archives [TNA]
PROB 11/699/19 (Mat[t]hew Wilson); PROB 11/1558/445 (Joseph Dawson)
Black history: https://webarchive.nationalarchives.gov.uk/ukgwa/20210803123212/https://livelb.nationalarchives.gov.uk/pathways/blackhistory/copyright.htm

Tyne and Wear Archives [T&WA]
DX104/1, 'A draught of Esq. Crowley's Works Beginning at The High Dam above Winlaton Mill and Ending at The River Tyne', *c.* 1728

University of Glasgow, Archives & Special Coll.
GB 247 MS Gen 501/27, letter from James Watt to James McGrigor, Glasgow, 30 Oct. 1784

West Yorkshire Archive Service [WYAS], Bradford
SpSt/5/5/2/15; 32D83, Records of Richard Hattersley

West Yorkshire Archive Service [WYAS], Leeds
WYL 899/222, Middleton Colliery ledger 1780–1; 6257/1-2, ledgers 1787–94
WYL76 (also listed as WYAS6755/10), Kirkstall Forge inventory, 1783
WYL463, John Brearley commonplace books (2 vols, 1758–62)

West Yorkshire Archive Service [WYAS], Wakefield
Deeds, CH 340 442 (1780); CN 408 542 (1783); CN 715 959 (1784)

Wiltshire and Swindon Archives
CRO 473/156 (Darby and Goldney agreement, 1713)

World of Wedgwood, Stoke-on-Trent, V&A Collection
Collect. Chem. Experiments 39-28405 (1771–6); 39-28406 (1768)

2 & 3 Anne, c. 4, Yorkshire (West Riding) Land Registry Act, 1703

Contemporary publications

Bailey's Northern Dir. (1781)
Baines, E., *History of the Cotton Manufacture in Great Britain* (Fisher, Fisher and Jackson, 1835) https://www.google.co.uk/books/edition/History_of_the_Cotton_Manufacture_in_Gre/CbY8AAAAcAAJ?hl=en&gbpv=1
The Book of English Trades (C. and J. Rivington for SPCK, 1827) https://archive.org/details/bookofenglishtraoounse/
Campbell, R., *The London Tradesman: being a Compendious View of all the Trades, Professions, Arts, both Liberal and Mechanic, now Practised in the Cities of London and Westminster* (T. Gardner, 1747; 3rd ed. 1757) https://archive.org/details/TheLondonTradesman/page/n7
Carlyle, T., *On Heroes, Hero-Worship and the Heroic in History* (1840) https://www.google.co.uk/books/edition/On_Heroes_Hero_worship_and_the_Heroic_in/kCo-AAAAYAAJ?hl=en
Clay (ed.), J. W., *Familiae Minorum Gentium* [by Joseph Hunter] (Harleian Soc., XXXVII, 1894) https://archive.org/details/familiaeminorumgo1hunt/page/n7/mode/2up
Dalton, J., *New System of Chemical Philosophy* (Manchester, 1808)
Defoe, D., *A Tour thro' the Whole Island of Great Britain, Divided into Circuits or Journies* (1724–7) https://www.visionofbritain.org.uk/travellers/Defoe
Defoe, D., *The Complete English Tradesman* (1726) https://www.gutenberg.org/files/14444/14444-h/14444-h.htm
Desaguliers, J. T., *A Course of Experimental Philosophy*, II (1744) https://www.google.co.uk/books/edition/_/g5MPAAAAQAAJ?hl=en&gbpv=1
Diderot, D. (ed.), *Encyclopédie ou Dictionnaire Raisonné des Sciences, des Arts et des Métiers* [Encyclopaedia, or a Systematic Dictionary of the Sciences, Arts, and Crafts] (1751–72)
Dudley, D., *D.D.'s Metallum Martis: or, iron made with pit-coale, sea-coale, etc. and with the same fuell to melt and fine imperfect mettals, and refine perfect mettals* (London, 1665) https://books.google.co.uk/books?id=qJkwooiujccC&printsec=frontcover&dq=metallum+martis+Dudley&hl=en&sa=X&ved=oahUKEwiYtqTV8K7pAhWEgVwKHV7GBG8Q6wEIKzAA#v=onepage&q=metallum%20martis%20Dudley&f=false
Fairbairn, W., *Treatise on Mills and Millwork. Part 1. On the Principles of Mechanism and on Prime Movers* (Longmans Green, 1861; 3rd ed., 1871)

https://www.google.co.uk/books/edition/Treatise_on_Mills_and_Millwork_On_the_pr/juxIAAAAMAAJ?hl=en&gbpv=1
Fortunes Made in Business (Sampson Lowe, 1884) vol. 1 https://www.google.co.uk/books/edition/Fortunes_Made_in_Business/PMZAAAAAIAAJ?hl=en&gbpv=1
Foster, J., *Pedigrees of the County Families of Yorkshire*, vol. 2 – the West Riding (Wilfred Head, 1874) https://archive.org/details/pedigreesofcount02fost/mode/2up
Furetière, A., *Dictionnaire Universel* (A. & R. Leers, The Hague and Rotterdam, 1690)
Henry, T., 'Considerations relative to the Nature of Wool, Silk, and Cotton, as Objects of the Art of Dying ... Together with some Observations on the Theory of Dying in general, and particularly the Turkey Red', *Memoirs of the Manchester Literary and Philosophical Soc.*, 3 (1790) 343–408 (read 20 Dec. 1786)
Hunter, J., *Hallamshire: A New Edition, with Additions by the Rev. Alfred Gatty* (Virtue & Co, [1875?])
Jars, G., *Voyages Métallurgiques, ou Recherches et Observations sur les Mines & Forges de fer ...* (G. Regnault, Lyon, 1774)
Jevons, W.S., *The Coal Question: An Inquiry Concerning the Progress of the Nation, and the Probable Exhaustion of Our Coal-Mines* (1865)
Lewis, W., *Proposals for Printing, by Subscription, the Philosophical Commerce of Arts* (London, 1748) https://www.google.co.uk/books/edition/_/JFtZAAAAcAAJ?hl=en&sa=X&ved=2ahUKEwiposKj_vjwAhVHzoUKHZg7BhoQre8FMA16BAgIEAY
Lewis, W., *The New Dispensatory* (5th ed., printed for C. Nourse in the Strand, 1785) https://iiif.wellcomecollection.org/pdf/b28776318
Lewis, W., *The Philosophical Commerce of Arts: Designed as an Attempt to Improve Arts, Trades, and Manufactures* (London, 1763–5) https://books.google.co.uk/books?id=UbpXAAAAYAAJ&printsec=frontcover&source=gbs_ge_summary_r&cad=0#v=onepage&q&f=false
Mortimer, T., *A Concise Account of the Rise, Progress and Present State of the Society for the Encouragement of Arts, Manufactures, and Commerce* (London, 1764) https://archive.org/details/b30787178
Percy, J., *Metallurgy: The Art of Extracting Metals from their Ores and Adapting them to Various Purposes of Manufacture*, Vol. 2 *Iron & Steel* (John Murray, 1864)
Plot, R., *Natural History of Staffordshire* (Oxford, 1686)
Rupp, T.L., 'On the Process of Bleaching with the Oxygenated Muriatic Acid; and a Description of a New Apparatus for Bleaching Cloths, with that Acid Dissolved in Water, without the Addition of Alkali', *Memoirs of the Manchester Literary and Philosophical Soc.*, 5 (1798), 298–313
Scrivenor, H., *A Comprehensive History of the Iron Trade* (Smith, Elder, 1841)

Smith, A., *An Inquiry into the Nature and Causes of the Wealth of Nations* (W. Straham & T. Cadell, 1776), I https://archive.org/details/inquiryintonatur01smit_o

Ure, A., *The Philosophy of Manufactures; or, an Exposition of the Scientific, Moral, and Commercial Economy of the Factory System of Great Britain* (Charles Knight, 1835)

Waller, T., *A General Description of All Trades* (T. Waller, London, 1747) https://books.google.co.uk/books?id=76mdQAAACAAJ&printsec=frontcover&source=gbs_ge_summary_r&cad=0#v=onepage&q&f=false

Yarranton, A., *England's Improvement by Sea and Land to Out-Do the Dutch without Fighting* [etc] (T. Parkhurst, 1677–81). Reissued in 1698: https://archive.org/details/englandsimpooyarr/page/n7/mode/2up

Theses and dissertations

Barraclough, K.C., 'The development of the early steel-making processes' (unpub. PhD, Univ. of Sheffield, 1981)

Bowman, J.F., 'The iron and steel industries of the Derwent valley: a historical archaeology' (unpub. PhD, Newcastle Univ., 2018)

Bruton, R.N., 'The Shropshire Enlightenment: a regional study of intellectual activity in the late eighteenth and early nineteenth centuries' (PhD thesis, Univ. of Birmingham, 2015) https://etheses.bham.ac.uk//id/eprint/5830/1/Bruton15PhD.pdf

Goss, J.C., 'A radical novelist in eighteenth-century England: Robert Bage on poverty, slavery and women' (MLitt thesis, Univ. of Birmingham, 2011) https://etheses.bham.ac.uk//id/eprint/3201/6/Goss_11_MLitt.pdf.

Hayman, R., 'The Shropshire wrought-iron industry *c*. 1600–1900: a study of technological change' (PhD thesis, Univ. of Birmingham, 2003) https://etheses.bham.ac.uk/id/eprint/248/1/Hayman04PhD_A1a.pdf

Hopkinson, G.G., 'The development of lead mining and of the coal and iron industries in north Derbyshire and south Yorkshire, 1700–1850' (PhD thesis, University of Sheffield, 1958) https://etheses.whiterose.ac.uk/21846/1/695348.pdf

Howes, A., 'Why innovation accelerated in Britain, 1651–1851' (unpub. PhD, King's College London, 2016)

Moher, J.G., 'The London millwrights and engineers, 1775–1825' (unpub. PhD, Royal Holloway London, 1988)

Books and articles

Allen, R.C., *The British Industrial Revolution in Global Perspective* (Cambridge U.P., 2009)

Anderson, R.G.W., 'Relations between Industry and Academe in Scotland, and the Case of Dyeing, 1760 to 1840', in L.L. Roberts and S. Werrett (ed.), *Compound Histories: Materials, Governance and Production, 1760–1840* (Brill, Cultural Dynamics of Science series, 2018), 333–53

Andrews, C.R., *The Story of Wortley Ironworks* (Milward, Nottingham, 1975)

Anstey, P.R., and A. Vanzo, 'Early Modern Experimental Philosophy', in J. Sytsma and W. Buckwalter (ed.), *A Companion to Experimental Philosophy* (Blackwell, 2016), 87–102 https://philarchive.org/archive/ANSEME

Ashton, T.S., *Iron and Steel in the Industrial Revolution* (Manchester U.P., 1963)

Ashworth, W.J., *The Industrial Revolution: the State, Knowledge and Global Trade* (Bloomsbury, 2017)

Awty, B. G., 'Charcoal Ironmasters of Cheshire and Lancashire, 1600–1785', *Trans Hist. Soc. Lancs & Cheshire*, CIX (1957), 71–121 https://www.hslc.org.uk/wp-content/uploads/2017/06/109-6-Awty.pdf

Baggs, A.P., Cox, D., et al., *A History of the County of Shropshire. XI. Telford* (1985) https://www.british-history.ac.uk/vch/salop/vol11

Barraclough, K.C., 'Wortley Top Forge: the Possibility of Early Steel Production', *Historical Metallurgy*, 11/2 (1977), 88–92

Berg, M., *The Age of Manufactures, 1700–1820: Industry, Innovation and Work in Britain* (Routledge, 1994)

Berg, M., and P. Hudson, 'Rehabilitating the Industrial Revolution', *Econ. Hist. Rev.*, XLV (1) (1992), 24–50

Berg, M., and P. Hudson, *Slavery, Capitalism and the Industrial Revolution* (Polity Press, 2023)

Berg, T. & P. (eds), *R. R. Angerstein's Illustrated Travel Diary 1753–1755: industry in England and Wales from a Swedish perspective* (translated by Torsten & Peter Berg, The Science Museum, London, 2001)

Biagioli, M., *Galileo, Courtier: the Practice of Science in the Culture of Absolutism* (Univ. of Chicago Press, 1993)

Bonnyman, B., 'Agrarian Patriotism and the Landed Interest: the Scottish "Society of Improvers in the Knowledge of Agriculture", 1723–1746', in K. Stapelbroek and J. Marjanen (ed.), *The Rise of Economic Societies in the Eighteenth Century* (Palgrave Macmillan, 2012), 26–51

Brearley, H., *Steel-Makers* (Longman Green, 1933)

Bruland, K., A. Gerritsen, P. Hudson and G. Riello, *Reinventing the Economic History of Industrialisation* (McGill-Queen's University Press, 2020)

Burnley, J., *The History of Wool and Wool-Combing* (1889)

Butler, A. E., B. F and H. M. (eds), *The Diary of Thomas Butler of Kirkstall Forge, Yorkshire, 1796–9* (privately printed, 1906)

Butler, R., *The History of Kirkstall Forge through Seven Centuries, 1200–1954* (Ebor Press, York, 1954)
Cantrell, J.A., and G. Cookson (ed.), *Henry Maudslay and the Pioneers of the Machine Age* (Tempus, 2002)
Cardwell, D.S.L., *The Organisation of Science in England* (Heinemann, 1972)
Cardwell, D.S.L., *The Fontana History of Technology* (Fontana, 1994)
Catling, H., *The Spinning Mule* (Lancashire Library, 1986)
Catling, H., 'The Development of the Spinning Mule', *Textile Hist.*, 9/1 (1978), 35–57
Chrimes, M.M. (ed.), *The Civil Engineering of Canals and Railways before 1850* (Routledge, 2017)
Clow, A. and N.L., *The Chemical Revolution: a Contribution to Social Technology* (Batchworth Press, 1952)
Coleman, D.C., 'Review of *Science and Technology* by Musson and Robinson', *Econ. Hist. Rev.*, 23/3 (1970), 575–6
Coleman, D.C., 'Textile Growth', in N.B. Harte and K.G. Ponting (ed.), *Textile History and Economic History: Essays in Honour of Miss Julia de Lacy Mann* (Manchester U.P., 1973), 1–21
Cookson, G., 'A City in Search of Yarn: The Journal of Edward Taylor of Norwich, 1817', *Textile Hist.*, 37/1 (2006), 38–51
Cookson, G., 'Hunslet Foundry and the Making of Industrial Leeds', *Yorkshire Archaeol. Jnl*, 93 (2021), 149–65 https://doi.org/10.1080/00844276.2021.1934245
Cookson, G., 'Quaker Networks and the Industrial Development of Darlington, 1780–1870', in J.F. Wilson and A. Popp (ed.), *Industrial Clusters and Regional Business Networks in England, 1750–1970* (Ashgate, 2003), 155–73
Cookson, G., *The Age of Machinery: Engineering the Industrial Revolution, 1770–1850* (Boydell, 2018)
Cookson, G., 'The Mechanization of Yorkshire Card-Making', *Textile Hist.*, 29/1 (1998), 41–61
Cookson, G., 'Wortley Forge: the Evolution of an Eighteenth-Century Ironworks', *Northern History*, 60/1 (2023), 52–73 https://doi.org/10.1080/0078172X.2022.2157364
Cookson, G. (ed.), *Victoria County History of Co. Durham* (University of London); *IV. Darlington* (2005); *V. Sunderland* (2015)
Cossons, N., *The BP Book of Industrial Archaeology* (David & Charles, 1993)
Cossons, N., (ed.), *Rees's Manufacturing Industry (1819–20)* (David & Charles, 1972)
Crossley, D. (ed.), *Water Power on the Sheffield Rivers* (Sheffield Trades Historical Society/University of Sheffield, 1989)
Crump, W.B. (ed.), *The Leeds Woollen Industry, 1780–1820* (Thoresby Soc. XXXII, 1931)

Evans, C., *The Labyrinth of Flames: Work and Social Conflict in Early Industrial Merthyr Tydfil* (Univ. of Wales Press, 1993)

Evans, C., O. Jackson and G. Rydén, 'Baltic Iron and the British Iron Industry in the Eighteenth Century', *Econ. Hist. Rev.*, 55/4 (2002), 642–65

Evans, C., and G. Rydén, *Baltic Iron in the Atlantic World in the Eighteenth Century* (Brill Academic Publishers, 2007)

Evans, C., and G. Rydén (ed.), *The Industrial Revolution in Iron: the Impact of British Coal Technology in Nineteenth-Century Europe* (Ashgate, 2005)

Farnie, D.A., and D.J. Jeremy (ed.), *The Fibre that Changed the World: the Cotton Industry in International Perspective, 1600–1990s* (Pasold, 2004)

Flinn, M.W. (ed.), *The Law Book of the Crowley Ironworks* (Surtees Soc., 167, 1952)

Flinn, M.W., *Men of Iron: the Crowleys in the Early Iron Industry* (Edinburgh U.P., 1962)

Gerbner, K., 'Slavery in the Quaker World', *Friends Jnl* (Sept. 2019) https://www.friendsjournal.org/slavery-in-the-quaker-world/

Gerhold, D., 'The Steel Industry in England, 1614–1740', in R.W. Hoyle (ed.), *Histories of People and Landscape: Essays on the Sheffield Region in Memory of David Hey* (Univ. of Hertfordshire Press, Studies in Regional and Local History, vol. 20, 2021), 65–86

Gibbs, F.W., 'William Lewis, M.B., F.R.S. (1708-1781)', *Annals of Science*, 8/2 (1952), 122–51

Gibbs, F.W., 'A Notebook of William Lewis and Alexander Chisholm', *Annals of Science*, 8/3 (1952), 202–20

Gibbs, F.W., 'William Lewis and Platina: Bicentenary of the Commercium Philosophico-Technicum', *Platinum Metals Rev.*, 7 (2) (1963), 66–9

Giles, C., and I.H. Goodall, *Yorkshire Textile Mills: The Buildings of the Yorkshire Textile Industry, 1770–1930* (HMSO, 1992)

Giles, C., and M. Williams (ed.), *Ditherington Mill and the Industrial Revolution* (Historic England, 2015)

Guest, J., *Historic Notices of Rotherham* (R. White, Worksop, 1879) https://www.google.co.uk/books/edition/Yorkshire_Historic_Notices_of_Rotherham/VpMWvgAACAAJ?hl=en&gbpv=1

Hadfield, C., and A.K. Skempton, *William Jessop, Engineer* (David and Charles, 1979)

Hahn, B., *Technology in the Industrial Revolution* (Cambridge U.P., 2020)

Harris, J.R., *Industrial Espionage and Technology Transfer: Britain and France in the Eighteenth Century* (Ashgate, 1998)

Hayman, R., 'The Cranage Brothers and Eighteenth-Century Forge Technology', *Hist. Metallurgy*, 38/2 (204), 113–20

Hayman, R., *Ironmaking: The History and Archaeology of the Iron Industry* (History Press, 2012)

Heaton, H., 'Benjamin Gott and the Industrial Revolution in Yorkshire', *Econ. Hist. Rev.*, 3/1 (1931)

Heaton, H., *The Yorkshire Woollen and Worsted Industry* (Clarendon, 1920) https://archive.org/details/yorkshirewoollen00heatuoft/page/288/mode/2up

Hey, D.G., *The Fiery Blades of Hallamshire: Sheffield and its Neighbourhood, 1660–1740* (Leicester U.P., 1991)

Hey, D.G., *The Village of Ecclesfield* (Advertiser Press, Huddersfield, 1968)

Hey, D.G., 'The South Yorkshire Steel Industry and the Industrial Revolution', *Northern Hist.*, 42/1 (2005), 91–6

Hey, D.G., (ed.), *The Militia Men of the Barnsley District, 1806: An Analysis of the Staincross Militia Returns* (University of Sheffield, 1998)

Hoover, H.C. and L.H. (ed. and trans.), *Georgius Agricola, De Re Metallica: Translated from the First Latin Edition of 1556* (Dover, New York, 1950) https://www.gutenberg.org/ebooks/38015

Humphries, J., and B. Schneider, 'Losing the Thread: a Response to Robert Allen', *Econ. Hist. Rev.*, 73/4 (2020), 1137–52

Hunt, L.B., 'The First Experiments on Platinum: Charles Wood's Samples from Spanish America', *Platinum Metals Rev.*, 29/4 (1985), 180–4

Hyde, C.K., 'The Iron Industry of the West Midlands in 1754: Observations from the Travel Account of Charles Wood', *West Midlands Studies*, 6 (1973), 39–40

Hyde, C.K., *Technological Change and the British Iron Industry, 1700–1870* (Princeton 1977)

Jacob, M.C., 'Mechanical Science on the Factory Floor: the Early Industrial Revolution in Leeds', *Hist. of Science*, 45/2 (2007), 197–221

Jacob, M.C., *The First Knowledge Economy: Human Capital and the European Economy, 1750–1850* (Cambridge University Press, 2014)

James, F.A.J.L., 'When Ben Met Mary: The Letters of Benjamin Thompson, Reichsgraf von Rumford, to Mary Temple, Viscountess Palmerston, 1793–1804', *Ambix*, 70/3 (2023), 207–328

James, J., *History of the Worsted Manufacture in England* (Bradford, 1857)

Jansson, M., and G. Rydén, 'Improving Swedish Steelmaking: Circulation and Localized Knowledge-Making in Early Modernity', *Technology and Culture*, 64/2 (April 2023), 515–42

Jenkins, D.T. and Ponting, K.G., *The British Wool Textile Industry, 1770–1914* (Pasold, 1982)

Jenkins, D. T., *The West Riding Wool Textile Industry, 1770–1835: a Study of Fixed Capital Formation* (Pasold, 1975)

John, A.H. (ed.), *Minutes Relating to Messrs. Samuel Walker & Co., Rotherham, iron founders and steel refiners, 1741–1829 and Messrs. Walkers, Parker & Co., lead manufacturers, 1788–1893* (Council for the Preservation of Business Archives, 1951) (Transcribed from a document in IMechE Archives: https://archivecat.imeche.org/records/BUS/17)

Jones, G., *Merchants to Multinationals: British Trading Companies in the Nineteenth and Twentieth Centuries* (Oxford U.P., 2000)

Kay, J., 'Adam Smith and the Pin Factory': https://www.johnkay.com/2019/12/18/adam-smith-and-the-pin-factory/

Kay, J., and M. King, *Radical Uncertainty: Decision-making for an Unknowable Future* (Bridge Street Press, 2020)

Kelly, M., J. Mokyr and C. Ó Gráda, 'Could Artisans have Caused the Industrial Revolution?', in Bruland et al., *Reinventing the Economic History of Industrialisation*, 25–43

Khan, Z., *Inventing Ideas: Patents, Prizes and the Knowledge Economy* (Oxford U.P. Scholarship Online, 2020)

King, P.W., 'The Vale Royal Company and Its Rivals', *Hist. Soc. Lancs and Cheshire*, 142 (1992), 1–18

Long, P.O., 'Of Mining, Smelting and Printing: Agricola's *De Re Metallica*', *Technology and Culture*, 44/1 (2003), 97–101

Long, P.O., 'The Openness of Knowledge: an Ideal and Its Context in Sixteenth-Century Writings on Mining and Metallurgy', *Technology and Culture*, 32/2 (1991), 318–55

Lyons, J.S., 'Powerloom Profitability and Steam Power Costs: Britain in the 1830s', *Explorations in Econ. Hist.*, 24 (1987), 392–408

Mann, J. de Lacy, *The Cloth Industry in the West of England, from 1660 to 1880* (Alan Sutton, 1987)

Mantoux, P., *The Industrial Revolution in the Eighteenth Century* (Jonathan Cape, 1948)

Marsden, B., *Watt's Perfect Engine: Steam and the Age of Invention* (Columbia U.P., New York, 2002)

Mathias, P., 'The Social Structure in the Eighteenth Century: a Calculation by Joseph Massie', *Econ. Hist. Rev.*, 10/1 (1957), 30–45

Maw, P., *Transport and the Industrial City: Manchester and the Industrial Age, 1750–1850* (Manchester U.P., 2013)

Milligan, E. H., *Biographical Dictionary of British Quakers in Commerce and Industry, 1775–1920* (Sessions Book Trust, 2007)

Mokyr, J., A. Sarid and K. van der Beek, 'The Wheels of Change: Technology Adoption, Millwrights and the [sic] Persistence in Britain's Industrialization', *Economic Jnl*, 132 (2022), 1894–1926

Mokyr, J., *The Enlightened Economy: Britain and the Industrial Revolution, 1700–1850* (Penguin, 2009)

Morley, C., *The Walkers of Masbrough: A Re-Examination* (self-published, 1996) [Copy in Rotherham Archives]

Morrell, J.B., 'Wissenschaft in Worstedopolis: Public Science in Bradford, 1800–1850', *British Jnl Hist. Science*, 18 (1985), 1–23

Morrison-Low, A.D., *Making Scientific Instruments in the Industrial Revolution* (Ashgate, 2007)

Mott, R.A., 'Early Ironmaking in the Sheffield Area', *Trans. Newcomen Soc.*, 27 (1949–51), 225-35

Mott, R.A., 'The Early History of Wortley Forges', *Bull. Historical Metallurgy Group*, 5/2 (1971), 63–70

Musson, A.E., and E. Robinson, *Science and Technology in the Industrial Revolution* (Manchester U.P., 1969)

Newman, J., and N. Pevsner, *The Buildings of England: Shropshire* (Yale U.P., 2006)

Newton, G.D., 'Farnley Smithies', *Yorkshire Archaeol. Jnl*, 88 (2016), 159–75

Newton, G.D., 'Wakefield, Its Woollen-Cloth Trade and Merchant Networks, 1558–1650' *Yorkshire Archaeol. Jnl*, 93 (2021), 129–48

Nigro, G. (ed.), *The Knowledge Economy: Innovation, Productivity and Economic Growth, 13th to 18th Century* (Datini Studies in Economic History, Firenze U.P., 2023)

Norris, J.M., 'The Struggle for Carron: Samuel Garbett and Charles Gascoigne', *Scottish Hist. Rev.*, 37, 124/2 (Oct. 1958), 136–45

Ó Gráda, C., 'Did Science Cause the Industrial Revolution?', *Jnl Econ. Literature*, 54/1 (2016), 224–39

Pacey, A.J., 'Emerging from the Museum: Joseph Dawson, Mineralogist, 1740–1813', *British Jnl Hist. Science*, 36/4 (2003), 455–69

Pole, W. (ed.), *The Life of Sir William Fairbairn, Bart.* (Longmans, Green & Co., 1877)

Porteous, J. Douglas, *Canal Ports: The Urban Achievement of the Canal Age* (Academic Press, 1977)

Raistrick, A., *Dynasty of Iron Founders: The Darbys and Coalbrookdale* (Longmans, Green, 1953)

Raistrick, A., *Quakers in Science and Industry* (David & Charles, 1968)

Raistrick, A. *Industrial Archaeology: an Historical Survey* (Paladin, 1973)

Raistrick, A., and E. Allen, 'The South Yorkshire Ironmasters, 1690–1750', *Econ. Hist. Rev.* 9/2 (1939), 168–85

Randall, A., *Before the Luddites: Custom, Community and Machinery in the English Woollen Industry* (Cambridge U.P., 1991)

Riden, P., 'The Output of the British Iron Industry before 1870', *Econ. Hist. Rev.*, 30/3 (1977), 442–59

Riden, P., *Gazetteer of Charcoal-fired Blast Furnaces in Great Britain in Use Since 1660* (Merton Priory Press, 2nd ed. 1993)

Riden, P., 'Navigation on the Don before 1726', forthcoming in *Yorkshire Archaeol. Jnl*, 96 (2024).

Riden, P., and J. G. Owen, *British Blast Furnace Statistics, 1790–1980* (Merton Priory Press, 1995)

Rimmer, W.G., 'Castle Foregate Flax Mill, Shrewsbury (1797–1886)', *Trans. Shropshire Archaeol. Soc.*, 56/1 (1957–8), 49–68

Rimmer, W.G., *Marshalls of Leeds, Flax-Spinners, 1788–1886* (Cambridge U.P., 1960)
Roberts, R., *The Classic Slum: Salford Life in the First Quarter of the Century* (Pelican, 1977)
Rogers, K.H., *Warp and Weft: the Somerset and Wiltshire Woollen Industry* (Barracuda Books, 1986)
Rolt, L.T.C., *The Mechanicals: Progress of a Profession* (Heinemann, 1967)
Rosenberg, N., *Inside the Black Box: Technology and Economics* (Cambridge U.P., 1982)
Ross, S., 'Scientist: the Story of a Word', *Annals of Science*, 18/2 (1962), 65–85
Rydén, G., 'Iron Production and the Household as a Production Unit in Nineteenth-Century Sweden', *Continuity and Change* 10/1 (1995), 69–104
Rydén, G., *Production and Work in the British Iron Trade in the Eighteenth Century: a Swedish Perspective* (Uppsala Papers in Economic History, 45, 1998)
Schubert, H.R., *History of the British Iron and Steel Industry, from c. 450 BC to AD 1775* (Routledge & Kegan Paul, 1957): https://archive.org/details/in.ernet.dli.2015.104051
Scott, E. Kilburn, 'Smeaton's Engine of 1767 at New River Head, London', *Trans Newcomen Soc.*, 19/1 (1938), 119–26
Secord, J.A., 'Knowledge in Transit', *Isis*, 95/4 (2004), 654–72
Sellars, M., 'Iron and Hardware', in W. Page (ed.), *Victoria History of the County of York*, II (1912)
Shapin, S., 'Paradigms Gone Wild', *London Review of Books*, 45/7 (30 March 2023)
Sheppard, F., and V. Belcher, 'The Deeds Registries of Yorkshire and Middlesex', *Jnl Soc. of Archivists*, 6/5 (1980), 274–86
Siegfried, R., *From Elements to Atoms: a History of Chemical Composition* (Trans. American Philosophical Soc, NS 92/4, 2002)
Sivin, N., 'William Lewis (1708–81) as a Chemist', *Chymia*, 8 (1962), 63–88
Skempton, A.W. (ed.), *John Smeaton FRS* (Thomas Telford Ltd, 1981)
Skempton, A.W., P. Cross-Rudkin and M.M. Chrimes, (ed.), *Biographical Dictionary of Civil Engineers*, I (Institution of Civil Engineers, 2002)
Smail, J., 'Manufacturer or Artisan? The Relationship between Economic and Cultural Change in the Early Stages of the Eighteenth-Century Industrialization', *Jnl of Social Hist.*, 25/4 (1992), 791–814
Smail, J., *Merchants, Markets and Manufacture: The English Wool Textile Industry in the Eighteenth Century* (Macmillan, 1999)
Smail, J., 'The Causes of Innovation in the Woollen and Worsted Industry of Eighteenth-Century Yorkshire', *Bus. Hist.*, 41/1 (1999), 1–15
Smail, J. (ed.), *Woollen Manufacturing in Yorkshire: the Memorandum Books of John Brearley, Cloth Frizzer at Wakefield, 1758–62* (YAS Record Series, CLV, 1999–2001)

Smiles, S. *Industrial Biography: Iron Workers and Tool Makers* (John Murray, 1863)

Smith, C.S., 'The Interaction of Science and Practice in the History of Metallurgy', *Technology and Culture*, 2/4 (1961), 357–67

Smith, R. Angus, *A Centenary of Science in Manchester* (Taylor and Francis, 1883) https://archive.org/details/acentenarysciencooolldgoog/page/n14/mode/2up

Soares, L. C., 'John Banks: an Independent and Itinerant Lecturer of Natural and Experimental Philosophy at the Threshold of the English Industrial Revolution', *Circumscribere*, 19 (2017), 18–33

Stembridge, P.K., *The Goldney Family: a Bristol Merchant Dynasty* (Bristol Record Society, XLIX, 1998) https://archive.org/details/bristol-record-society-49

Styles, J., 'The Rise and Fall of the Spinning Jenny: Domestic Mechanisation in Eighteenth-Century Cotton Spinning', *Textile Hist.*, 51/2 (2020), 195–236

Tate, W.E., 'The Five English District Statutory Registries of Deeds', *Historical Research*, XX/60 (May 1944), 97–105

Taylor, L., and S. Levon, *John Smeaton and the Calder Navigation, with the transcription of John Smeaton's Journal 1760–1763 detailing the day-to-day work on the Navigation* (Wakefield Historical Publications, 2021)

Thompson, E.P., *The Making of the English Working Class* (Pelican Books, 1968)

Tweedale, G., *Steel City: Entrepreneurship, Strategy and Technology in Sheffield, 1743–1993* (Clarendon Press, 1995)

Unsworth, R., *Leeds: Cradle of Innovation* (Leeds Sustainable Development Group, 2018)

Unwin, R.W., 'Leeds Becomes a Transport Centre', in D. Fraser (ed.), *A History of Modern Leeds* (Manchester U.P., 1980), 113–41

Wakefield, A., 'Butterfield's Nightmare: the History of Science as Disney History', *History and Technology*, 30/3 (2014), 232–51

Willan, T.S., 'Yorkshire River Navigation, 1600–1750', *Geography*, 22/3 (Sept. 1937), 189–99

Wilson, R.G., *Gentlemen Merchants: The Merchant Community in Leeds, 1700–1830* (Manchester U.P., 1971).

Wilson, R.G., 'The Supremacy of the Yorkshire Cloth Industry in the Eighteenth Century', in N.B. Harte and K.G. Ponting (ed.), *Textile History and Economic History: Essays in Honour of Miss Julia de Lacy Mann* (Manchester U.P., 1973), 225–46

Wilson, R.G., 'The Textile Industry', in C. Rawcliffe and R. Wilson (ed.), *Norwich since 1550* (Hambledon & London, 2004)

Wilson, R.G., 'Georgian Leeds', in D. Fraser (ed.), *A History of Modern Leeds* (Manchester U.P., 1980), 24–43

Wootton, D., *The Invention of Science: a New History of the Scientific Revolution* (Penguin, 2016)

Online sources

John Kanefsky, Early Engines Database: https://coalpitheath.org.uk/engines/index
Malcolm Dick: https://www.revolutionaryplayers.org.uk
Quakers in the World: https://www.quakersintheworld.org/quakers-in-action/11/Anti-Slavery
YAHS Industrial History Online: https://www.industrialhistoryonline.co.uk/yiho/index.php

Index

Aberdeen   125, 180–1
Académie des Sciences   76
Agricola, Georgius   71
Albion Mill (Southwark)   107
Allen, Robert   202–3
Anderson, John   181
Anderson, Robert G.W.   180
Angerstein, Reinhold Rücker   20, 24, 30–5, 43, 47, 57, 61, 63, 72, 144, 188, 192, 196–7, 210
apprenticeship and apprentices   26, 99–101, 107–8, 110, 131, 148, 151, 159, 179, 188fn
    premium apprenticeship   18, 101, 107–8, 110
Arkwright, Richard   28, 34, 89, 91 and fn, 92, 118, 122, 140, 167, 176, 198, 201, 204, 208, 210
armaments   2, 36, 39, 51, 207
Armley (Leeds)   19
artisans *see* social class
Ashton, T.S.   43–4, 58
Aston, Birmingham   32
Atherton, Peter   210
Atlantic trade *see* North American colonies
Austen, Jane   101
Austhorpe (Leeds)   86–7
Aydon & Elwell   131
Ayrshire   126

Bacon, Anthony   47
Bacon, Francis   67–9, 70–1
Bage, Charles   118, 119–20, 127, 167, 210
Bage, Robert   118
Baines, Edward   199
Bank Furnace (Bretton)   158
banks and banking   2, 9, 15, 44, 158, 166
Banks, John   130–2, 134, 210
Banks, Sir Joseph   121–2, 139

Barnes, Thomas   122–3, 211
Barnsley   x, 5, 32, 78, 144, 150, 208
Baylies, Thomas   44
Benyon, Thomas and Benjamin   118, 211
Berg, Maxine   209
Bergscollegium (Swedish Board of Mines)   30
Bersham Foundry (Wrexham)   45, 164
Berthollet, Claude Louis   70, 90, 124
Bertram, Wilhelm   21, 24, 33, 176, 211
Berzelius, Jöns Jacob   67
Bierley ironworks   60
Birkbeck, George   181
Birkenshaw iron foundry   60
Birmingham   x, 32, 33, 40, 43, 90, 82, 114–17, 126–7, 138, 176, 179, 181, 195, 196
Black Country   x, 21, 28, 40, 143, 150
Black, Joseph   89–90, 116, 120, 137, 138, 178–81, 183, 211
Blackhall (Tyneside)   21, 24, 33, 176, 211
Board of Trustees for Fishing, Manufactures and Improvements (Board of Manufactures)   82–3, 142, 143, 177–9, 206
Boerhaave, Herman   180
Booth, John   162 and fn
Boswell, James   139
Boulton & Watt   17fn, 39, 55, 91–2, 107, 124, 126–8, 132, 139, 140–1, 164, 167, 170, 179, 184, 199, 201–2
    Soho Foundry, Smethwick   92, 126–7, 128, 181
Boulton, Matthew   90, 115, 117, 126, 167, 179, 211
Boulton, Matthew jun.   92, 127
Bowling ironworks   60
Boyle, Robert   71, 83, 133
Bradford   60, 120, 164
Bradford archives (WYAS)   95

Bradford-on-Avon   17
Brandling family   176, 211
Brass-founding   39–40
Brearley, John   12, 15–16, 20, 145
Brindley, James   143
Bristol   21, 28, 39–40, 42, 44, 143, 146, 165
Broadbent, Joseph   158, 160–1, 209, 212
Brooke, Thomas   35
Brownrigg, William   47, 81, 180
bruk (Swedish)   32, 186
Bruton, Roger   119–20, 134
Burdon, Rowland   212
business management, business premises *see* systems of production
Butler & Beecroft   59, 60, 104, 148, 190–1
See also Kirkstall Forge

Calder Iron Works (Airdrie)   183
Calley, John   138
Campbell, R.   109–10
canals and river navigations   9–11, 17, 22, 28, 85, 90, 143–5, 176, 207
   Aire and Calder Navigation   10–11, 22, 85, 87, 136fn, 144
   Bradford Canal   60, 144, 164, 169
   Calder and Hebble Navigation   87
   Don Navigation   147, 162
   Glamorgan Canal   168
   Kennet and Avon Canal   17fn
   Leeds and Liverpool Canal   144
   Shropshire Canal   166
   Somerset Coal Canal   143, 201
   See also rivers and watercourses
Cannon Hall (Cawthorne)   157
Cardwell, Donald   67, 121–2, 134
Carlyle, Thomas   7, 171
Carnot, Nicolas Léonard Sadi   86
Carr, William   151, 167, 212
Carron Company, Falkirk   45–6, 48, 52, 55, 58fn, 86, 90, 126, 147, 150, 164, 169–70, 179–80, 182–3, 195
Cartwright, Edmund   198
Chalmers, Dr Thomas   131–2
Chambers of Commerce   181
Chandos, Duke of   134
Chapel Furnace (Chapeltown, Sheffield)   51, 155
charcoal production and usage   12, 20, 31–9, 42, 43–5, 48–52, 55, 57–8, 60, 72, 153–5, 185–6, 189, 192, 209

chemical production   67, 70, 73–4, 177–82, 204
chemistry   46, 48–9, 63–7, 69–70, 73–8, 81–2, 88, 90, 93, 120–9, 133, 141, 166, 169, 176–83, 189, 192–3, 197, 204, 206
   alchemy   70–1, 177, 206
   chemical societies   181
   See also periodic table of elements
Chesterfield   51, 80, 155
Chisholm, Alexander   73, 77–81, 133, 188
Chowbent   47fn, 131, 151fn
Clifton (Cumberland)   35
Clockmaking   89, 106
Clow, Archibald and Nan L.   66, 67, 126, 177, 179
Clyde Iron Works   183–4
Coalbrookdale   33, 35, 39–42, 44–5, 47–8, 72, 80–1, 144fn, 146, 155, 165, 166, 176, 186–7, 200
   See also Darby family
coal-mining   x, 12, 25, 31–2, 45, 61, 78, 98, 104, 144, 186
Coalport   144fn, 166, 186
Cockshutt family of Wortley Forge   4–5, 54–5, 57, 59, 78, 80, 161, 164, 169, 170, 205, 212
   Cockshutt, Edward   161
   Cockshutt, James   54–6, 57, 80, 87, 168, 212
   Cockshutt, John I   60, 149, 156–9, 161, 212
   Cockshutt, John II   79–80, 198, 208, 212
coking   41, 54, 226
Colne Bridge ironworks (Mirfield)   148, 158
commercial change   9, 17, 19, 22, 25, 44–5, 118, 123, 132, 168, 179–80, 184–5, 189–90
community   9, 25–6, 59, 104–5, 123, 158, 173, 184–93, 194, 198, 204–5
Company in the North   21, 216
Company of Scotland *see* Darien Company
Cooper, Dr Thomas   124
Cope, Thomas   157–8
Cornwall   x, 12, 124, 126, 128, 201
Cort, Henry   34, 37–9, 43, 47–50, 52–7, 67, 141, 183, 192, 198, 204, 212, 227

INDEX

Cossons, Neil 41
Cotton, William, and Cotton family 153-4, 159
Cranage method and brothers (Thomas and George) 47-9, 72, 81, 182-3, 192, 212
Crawshaw (or Crawshay), John 161-4
Crawshay, Richard 53-5, 57, 60, 120, 168, 192, 205, 208, 212
Crompton, Samuel 140-1, 170, 198, 213
Cross Street chapel, Manchester 122, 207; and Appendix, 210-25
Crowley, Ambrose (1635-1720) 21
Crowley, Ambrose (1658-1713) 20-5, 33, 143, 146, 153, 176, 185-8, 190, 207, 213
Cullen, William 178
Cumberland x, 12, 22, 35, 47, 121, 150, 176, 180
Cumbria 10
Cupola Company *see* Hunslet Foundry
cutlers 50, 87, 147-8, 185
Cyfarthfa ironworks (Merthyr Tydfil) 47, 53-5, 146, 168, 192

Dalton, John 67-8, 121-3, 125, 129, 133, 134fn, 141, 180, 193, 204, 213
Darby Company 39-50, 57, 80-1, 156, 164-6, 169, 185-7, 192, 205
  Darby, Abraham I 20, 38, 39-44, 50, 61, 142-3, 146, 182, 213
  Darby, Abraham II 33, 34, 44-6, 50, 51, 81, 200, 213
  Darby, Abraham III 213
  Darby, Mary (wife of Darby I) 44
  Darby, Mary (daughter of Darby I) 44
Darien Company 2, 177
Darwin, Dr Erasmus 115, 117-19, 121, 133, 179-80, 206, 213
Dawson, Joseph 93, 115, 122, 164, 169, 179-80, 183, 192, 205, 207, 208, 214
deeds registry 9, 22, 23
Defoe, Daniel 14, 24, 37, 171
Derby 35
Derbyshire 12
Derwent Valley 21, 33, 154, 186
Desaguliers, John Theophilus 4, 7, 77, 90, 130, 134-9, 200, 214
Dickens, Charles 171-2
Dickin, Thomas, and Dickin family 153-4

dissenting academies 116, 175
  Daventry 115, 120, 169
  Kendal 120, 131
  Northampton 179
  Warrington 120-1
  See also Manchester Academy
Diderot, Denis 195
Ditherington Mill (Shrewsbury) 118-20, 127, 167
division of labour *see* systems of production
Drabble, William and Joseph 152, 185fn, 214
Dudley (Worcestershire) x, 40, 41, 138, 139fn
Dudley, Dud 22, 40, 176, 214
Duke of Norfolk's ironworks 161
Dundonald, Archibald Cochrane, 9th Earl of 66
Dunfermline 125
dyeing and bleaching 11, 69-70, 75-6, 90, 122-5, 141, 144, 178-9, 181, 182, 192-3, 204

East Anglia 12, 17, 136 fn, 203
economics of production 22-4, 32, 38-9, 42-3, 51-3, 70, 72, 75, 95, 98, 142, 178, 199-201, 203
economists 4, 7, 22, 81fn, 101, 195-7
Eddystone Lighthouse 85
Edinburgh 178, 180, 181
  See also universities
education 7, 18, 22, 26, 30, 40, 62-3, 69, 80, 84, 95-9, 101-4, 111, 113, 116, 118-19, 120-2, 124, 128-9, 132, 136-7, 140, 155, 161, 169, 174-5, 179-82, 204-6
  education of women 118-19
  See also dissenting academies; Manchester; universities
Egremont, Cumberland 47
Elam family 19-20, 24, 60, 101, 170, 204-5, 209
  Elam, Emmanuel 19, 214
  Elam, Gervase 19, 214
  Elam, John 19, 214
Enlightenment, The 5
Eno, Brian 172-3
Evans, Chris 36-7, 65-6, 146, 153, 186, 188-9, 197
Ewart, Peter 111

Eyres, Wigglesworth & Co.   59

factory building   8, 60, 93, 98, 102, 105–6, 108–9, 111, 118–20, 125, 127–8, 174, 190, 201
Fairbairn, Peter   108, 109, 151, 179
Fairbairn, William   107, 108, 110–12, 179, 201, 214
Fell, John I and John II   154, 155, 159, 160, 161
Ferguson, James   133
file-cutting   32, 78, 89fn, 148, 149
Foley and associates, iron combines (west Midlands)   61, 153, 185
Ford, Richard   44–5, 215
Forest of Dean   35
France   5, 31, 43, 57, 66, 76, 77, 86, 116, 132, 146, 147fn, 195
Franklin, Benjamin   121
freemasonry   103, 117, 175; and Appendix, 210–25

Garbett, Samuel   176, 182–3, 215
gas engineering   92, 123–8, 132, 135, 139, 141, 193, 204, 207
Gascoigne, Charles   55, 164, 182–3, 215
George III   74, 180
George, Watkin   54, 168
Gerbner, Katharine   165
Germany   21, 22, 24, 30, 32, 66, 73, 81, 147
Gibbs, F.W.   77–81
Glasgow   89–90, 93–4, 115, 124, 137, 169, 178, 179–182
global commodity chain   36–7, 143, 187, 191–2
Goldney, Thomas II   44, 165
Goldney, Thomas III   44–5, 47–8, 215
Goodrich, Simon   188
Gothard, Timothy   151–2, 167, 215
   See also Salt & Gothard
Gott, Benjamin   4fn, 26, 60, 69, 128, 170, 190, 205, 215
Greenwich (London)   21
Grenoside (Ecclesfield)   5, 102, 149, 161–2
Griffiths, Gabriel   46

Hadley, John   136 and fn, 137
Hales, Dr Stephen   74, 77, 83, 114
Halifax   x, 11, 18–19
Hanbury, John   21, 32 and fn, 227

Hargreaves, James   140–1, 203
Hattersley, Richard   61, 95, 97, 99, 104, 108, 151–2, 184–5, 190, 216
Hauksbee, Francis   134
Hayford, Denis [or Heyford]   21–2, 47fn, 154, 176, 216
Hayman, Richard   46, 48, 185, 188, 189
Hazledine, William   216
Henry, Thomas   121–5, 128–9, 179, 180–1, 216
Henry, William   88–9, 121, 123, 127–8, 129, 133, 139, 206, 217
Hewes, Thomas C.   111
Hey, William   88, 114
Hodgkinson, Eaton   129, 217
Hudson, Pat   209
Huguenot refugees see migration
Hull   10–11, 14, 144
Humphries, Jane   202–3
Hunslet Foundry (Leeds)   4–5, 95–8, 101, 104, 108, 151–2, 164, 167, 215, 222
Huntsman, Benjamin   32, 42fn, 50, 89, 147, 217, 226
Huthwaite (Silkstone)   159
Hutton, James   178
Hyde, Charles K.   42, 51

industrial enlightenment   2, 103, 134
instrument-making   66, 85, 88–91, 132, 134, 137, 138, 147, 150, 169fn,
iron   30–61
   cast iron   33, 35, 38, 39–43, 45, 46, 51, 54–6, 58, 60, 75, 80, 92, 107–8, 111, 112, 127, 131, 151–2, 161–2, 168, 176, 200, 226
   iron-making   3, 20, 32–3, 62–5, 73–82
   ironstone mining   20, 25, 41, 61, 70, 72, 78, 148, 153–4, 155, 164, 177, 183–4
   iron trade   5, 15–18, 20–1, 24, 25, 36–7, 53, 142–3
   ironworkers   13, 61, 65, 140, 146–7, 184–91
   iron-working businesses   4–5, 21, 152–70, 184–91
      See also Bierley, Birkenshaw, Bowling, Clyde, Colne Bridge, Cyfarthfa, Duke of Norfolk's, Shelf ironworks; Bank, Chapel, Rockley furnaces; Darby Co.; Kirkstall Forge; Low Moor Co.

production volume   50–2, 184
products   36, 41–2, 51, 60, 150, 185–7, 189, 190
wrought (malleable) iron   37–8, 39, 41, 43, 49, 55, 60, 92, 112, 162, 226–7
iron technologies
   blast furnace   32, 38–41, 43, 45, 54, 182, 226
   blowing engines   46, 52, 168, 182
   charcoal-burning furnaces   12, 35, 49–52, 185
   coke-fired furnaces   39–40, 45, 49–52, 58–60
   cupola furnace   58–9, 104, 152, 164, 193, 199, 226
      See also Hunslet Foundry
   forging and smelting techniques   39–43
   iron-smelting   31–2, 34–5
   puddling and rolling   3, 6–7, 33, 34, 37–8, 43, 47–9, 52–7, 59, 65–6, 82, 135, 166, 170, 183–4, 188–9, 192, 201, 207, 227
   reverboratory (air) furnace   30, 38–41, 47, 48–9, 58, 161, 182, 227
   stamping and potting   47, 51, 53, 57, 80, 189, 201, 227
Italy   57

Jackson, Owen   153
Jars, Gabriel   31
Jellicoe, Adam   49
Jellicoe, Samuel   49
Jessop, William   87, 217
Jones, Geoffrey   18
Joule, James Prescott   129
Jubb, John (I)   102, 107–8, 151–2, 217–18

Kandinsky, Wassily W.   172
Kay, John (1704–80/81)   34
Kay, John (economist, b. 1948)   195–6
Keighley   x, 61, 95, 97, 104, 108, 144, 146, 151, 184, 190, 193, 198
Keir, James   124
Kendal   10
Kennedy, John see M'Connel & Kennedy
Kirk Mill (Chipping)   167
Kirkstall Forge (Leeds)   59–60, 104, 107, 144, 148, 151–2, 158, 160, 164, 190–1
knowledge   3–6, 8–29, 69–72, 82–3, 93–4, 109–10, 122, 124, 129–30, 134–5, 146, 170, 189–93, 206–8

Lancashire   8, 11, 20, 35, 45, 70, 78, 80, 86–7, 124–5, 151, 154, 179
Landes, David S.   99
language   5, 63, 64, 66, 72, 107, 192
Lavoisier, Antoine   66–7, 115, 121, 125
Lead Hill (Lanark)   77
learned societies see Literary and Philosophical Societies; Lunar Society; Royal Society; Royal Society of Arts
Lee, George Augustus   127–8, 141, 207, 218
Leeds   x, 4, 11, 17–20, 24–5, 28, 35, 51, 59–60, 69–70, 85–8, 101–2, 107–9, 114–15, 118, 120, 128, 131, 144, 146, 147, 150–2, 169fn, 170, 176, 183, 190, 193, 198, 200–1, 203, 204–5
Lewis, William   48, 63, 64, 66, 73–83, 88–9, 93, 113, 114, 133–4, 179, 192, 197, 198, 208, 218
   Lewis archive   74, 77–82
   table of affinities   66, 73, 75, 76
Liège (Belgium)   21, 24, 32
Lister, William   167
Literary and Philosophical Societies   7, 93–4, 114–29, 130, 141, 146, 175, 181–2, 203
   Bradford Literary and Philosophical Society   120, 169
   Derby Philosophical Society   116–19, 121
   Glasgow Philosophical Society   181
   Leeds Philosophical and Literary Society   88, 114, 120, 203
   Literary Society of Glasgow   181
   Manchester Literary and Philosophical Society   119, 120–9, 131, 141, 179–80, 182, 192–3, 203, 206, 207
Liverpool   14, 25
London   11, 20, 24, 109–11, 138, 145
Long, Pamela O.   70–1
Low Moor Co. (Bradford)   60, 93, 164, 169, 183, 192, 208
Lunar Society   115–16, 117, 119, 121, 126, 128, 139, 141

machinery and mechanization   3, 12, 14, 56, 58–9, 62–3, 95, 98, 103–12, 150, 170, 179–80, 184–5, 189, 190, 193, 198–200, 205
   machine-breaking   202

machinery and mechanization (*continued*)
  machine tools   3, 60, 92–3, 105, 108–9, 111, 126, 150, 185fn, 198, 200
Maclea, Charles Gascoigne   183
Manchester   10, 14, 25, 28, 131
  College of Arts and Sciences   123–4
  Manchester Academy (New College)   121, 123, 131, 180–1
Marshall, John   4fn, 60, 69–70, 118, 170, 218–19
Massachusetts   74
Massie, Joseph   101
mathematics   62, 70, 76, 84–6, 89, 93, 107, 110, 112, 121, 130–1, 134–8, 140, 175, 193, 200, 205–7
Maw, Peter   10
McGrigor, James   182
M'Connel & Kennedy   131, 179, 218
  Kennedy, John   131
Mechanics' Institutes   181; and Appendix, 210–25
merchants   9–25, 59–60, 100–2, 158, 167–8, 170, 192
Merthyr Tydfil   45, 47, 53, 189
metallurgy   34, 63, 66, 67, 69–71, 81, 180, 183
Middleton Colliery and waggon-way   59, 86, 95–8, 104, 152, 176, 211
Midlands (English)   8, 11–12, 22, 28, 35, 41–2, 46, 61, 67, 78, 92, 117, 119, 122, 143, 148, 150, 153, 176, 185, 188, 206
migration, migrants   7, 24, 25, 27–8, 32, 59, 61, 100, 142, 145–52, 171, 179, 191–2, 193, 194
  emigration   146–7
  Protestant refugees   19, 24, 146
Mill Hill chapel (Leeds)   114–15, 169fn
millwrights   7, 85, 98, 100, 102, 105–12, 126, 130, 137, 139, 140, 145, 151, 192
Moher, James G.   110
Mokyr, Joel   2, 27, 106fn, 134
Murdock, William   126–8, 139–40, 170, 179, 184, 202, 219
Murray, Matthew   69–70, 91–3, 99, 103, 104, 108, 147, 151, 167, 170, 183, 185fn, 202, 219
Mushet, David   93–4, 115, 122, 170, 183–4, 207, 219
Musson, Albert E., and Eric Robinson   3fn, 70fn, 111, 118

nail-making and -trading   21, 25, 32–3, 56, 104, 147–51, 154 157, 158, 160–2, 167, 185, 187, 190, 195, 204, 227
Navy Board *see* Royal Navy
Netherlands   18, 20, 21–2, 24, 32, 103, 136fn, 137, 146–7
Newcastle upon Tyne   x, 21, 43, 120, 143, 176
Newcomen, Thomas   4, 17fn, 34, 39, 41fn, 45, 51, 59, 87, 89, 91, 137–40, 199, 201, 219
Newton, George   10–11
Newton, Isaac   83, 113, 132, 134–5
Norfolk   x, 17
North American colonies   3, 9, 18–20, 24, 31, 36, 73–4, 116, 142–3, 165, 170, 205, 207
Northumberland, Duke of   74
Norwich   12, 13, 17
Nottingham   32

Oakes of Bury St Edmunds   202
Oates, James   159–60
organizational models in industry *see* systems
Öregrund (Sweden)   32, 33, 36, 43
Owen, Samuel   183, 220

Pacey, Arnold J.   113
Paine, Thomas   116, 118
Panama   177
Papin, Denis   4, 138–9, 219
partnership   44, 54–5, 152–70
patenting   37, 38, 40, 42, 45, 47–50, 53, 55, 58fn, 80–1, 90, 92, 120fn, 124, 126, 127, 132, 136fn, 139, 140, 167, 183, 201–2
Paul, Lewis   24, 167, 208, 220
Penrith   131
Percival, Thomas   121–4, 179–80, 206, 220
Percy, John   53–4
periodic table of elements   6, 66–7, 68, 73
  See also chemistry
Peter I, Tsar of all Russia (Peter the Great)   135
Philadelphia (Penn., U.S.A.)   19, 181
Philips & Lee   127–8
  Philips family   127
philosophical societies *see* Literary and Philosophical Societies; science: public science

INDEX

platinum 73, 75, 80–1
Plot, Robert 38, 49
Plymouth 14, 85
Pole, William 110–11
policy-making 37, 48, 142, 178–9, 206
Pontypool 21, 32, 54
Poor Law 99–100, 147, 149
ports, dockyards and harbours 9, 12, 28, 85, 217, 221
potash 73–4, 75, 82, 83, 178
Prestonpans (East Lothian) 179–80
Priestley, Joseph 64, 66, 88, 114–16, 120–1, 125, 132–3, 220
Proctor, Thomas 40

Quakers 19, 20fn, 39, 44–5, 47–8, 50, 121, 165–6, 175, 207

Raistrick, Arthur 48
Réaumur, René Antoine Ferchault de 77
Redruth (Cornwall) 126
Rees, Abraham 71
regions 7, 8–14, 17–18, 27–8, 41–2, 142–5
Reinhard, Johann Jacob 81
Renishaw 51
religious affiliations 103, 105, 120, 129, 164, 166, 169, 175, 179, 204, 206; and Appendix, 210–25
See also Quakers
Rennie, John 107, 143, 179, 221
Reynolds, Richard 47–8, 72, 81, 221
Reynolds, Susan 5 and fn
Reynolds, William 120, 165–6, 176, 207, 221
Richmond, Surrey 75
Riden, Philip 34–5, 50–1
Rinman, Sven 31, 34, 72
rivers and watercourses 22, 87, 105–6, 143, 187
  Derwent 21, 33, 154, 186
  Don 52, 147, 148–9, 201
  Humber 10, 19, 22, 147fn
  Severn 21, 35, 42, 143–4, 166, 186–7
  Thames 21
  Trent 147fn
  See also canals and river navigations
roads see transport infrastructure
Roberts, Richard 208
Roberts, Robert 101
Robison, John 89–90, 221
Rockley Furnace (near Wortley) 154

Roebuck, John jun. 46, 90, 179–80, 182, 183, 206, 221–2
Rolt, L.T.C. (Tom) 138–9
Rosenberg, Nathan 4
Rotheram, Caleb 131
Rotherham x, 39, 51, 147, 148–9, 151, 162
  See also Walker ironmasters
Royal Institution 122fn
Royal Navy 32, 36, 162, 186–7
  Navy Board 48–9
Royal Society 71, 73, 81, 83, 114–15, 120, 121–2, 129, 132, 134, 138, 175, 206
Royal Society of Arts (Society for the Encouragement of Arts, Manufactures and Commerce; from 1908, RSA) 64, 73, 80, 83, 114, 124, 173 fn, 175
Rupp, Theophilus Lewis 125, 222
Russia 20, 31, 34, 36, 43, 55, 103, 143, 146, 147, 163fn, 172–3, 182
Rydén, Göran 36–7, 43, 153, 186, 197

Saint, Samuel 162
Salford 101, 127
Salt & Gothard 96, 98
Salt, Titus 151, 222
Savery, Thomas 4, 138, 219, 225
Scenius 2, 172–3, 185, 193
Scheele, Carl Wilhelm 66
Schlüter, Christoph Andreas 71
Schneider, Ben 202–3
Schröder, Samuel 31, 33, 37, 187, 192
Schubert, H. R. 52, 78, 80
science 2–4, 7, 62–94
  enlightenment science 2, 62, 89, 105, 123, 145
  public science 7, 88, 105, 114, 118, 122, 130–9
  science and arts 64–5, 73–8, 80, 83, 122–4, 181
  science and industry 31, 33–4, 62–6, 105, 114–16, 128–9, 134, 141, 161, 177–84, 207
  science and technology 6, 62, 84–94, 122, 144, 204
  scientist, meaning of 64, 66, 72, 139
  See also chemistry; mathematics
Scotland 2, 7, 12, 20, 28, 30, 45–6, 52, 67, 82–3, 113, 120, 125, 128, 142–3, 151, 174, 176, 177–84, 204, 206
  Act of Union (1707) 2, 177

Scottish Enlightenment 89, 116
Scottish Society of Improvers in the
   Knowledge of Agriculture 178
Seacroft foundry (Leeds) 51, 59, 86
Seamer (Ayton) Forge 78, 148
Sheffield x, 32, 35, 43, 50, 51, 59, 80, 87,
   100, 104, 120, 147–52, 155, 158, 179,
   209
Shelf ironworks 60, 131
Shipley, William 83
Shrewsbury 11, 13, 28, 40, 118, 203
Shropshire 12, 41, 45, 47, 119–20, 143,
   146, 150, 153, 166, 182, 188–9
   Shropshire Enlightenment 119
Simpson, William 154
Sivin, Nathan 73, 76
slavery 20, 165–6, 207, 209
Smail, John 14, 15, 17, 18, 203
Smeaton, John 46, 52, 59, 64, 84–8,
   93, 110, 112, 113, 115, 135–6, 176,
   182–3, 200, 206, 222
Smeatonian Society of Civil
   Engineers 106
Smiles, Samuel 50, 172, 196
Smith, Adam 7, 195–7
Smith, William 222
social class 2, 7, 25–7, 62, 65, 84, 98–105,
   126, 139, 140–1, 150, 159, 166, 175,
   205
   artisans 26, 64–5, 88, 93–4, 95–8, 100,
      101, 103–4, 107–8, 147fn, 175,
      196, 205–6
Society for the Encouragement of Arts,
   Manufactures and Commerce *see*
   Royal Society of Arts
Society of Friends *see* Quakers
Soho Foundry (Smethwick, Birmingham)
   see Boulton & Watt
Sorocold, George 136 and fn, 137
sources and approaches 3–6, 15–16, 74
   and 77–82, 84, 95–8, 104, 105, 142,
   184–5 (& plates 12–14), 203
Spain 36, 147fn
Spencer iron syndicates and partners 21,
   43, 59, 61, 80, 144, 148, 150, 153–61,
   164, 169, 175, 176, 185, 190, 193, 222
   sites *see* Bank; Chapel; Colne Bridge;
      Duke of Norfolk's; Kirkstall;
      Rockley; Seacroft; Seamer;
      Staveley; Thurgoland; Wortley
Spencer, Edward 156, 161
Spencer, John I 153–4

Spencer, John III 156
Spencer, John IV 155, 156, 161
Spencer, William 149, 155–61
Staffordshire 12
Stahl, Georg Ernst 66
Staveley ironworks (Chesterfield) 51, 80,
   155
steam engines 12, 17, 28, 31–2, 39, 45,
   51–2, 58, 60, 85–7, 89, 90–4, 106,
   108, 115, 124, 126, 131, 135, 137–9,
   144, 147, 179, 185, 200–2
   See also Watt, James
steel-making and products 21, 24, 31–3,
   37, 43, 50, 75, 78, 89, 147, 152, 162,
   183, 186, 226
Stephenson, George 102, 205
Stephenson, Robert 102
Stevenson, Robert Louis 113
Stirling, James 134
Stockton and Darlington railway 143
Stourbridge 21
Strutt, Jedediah and Elizabeth 222–3
Strutt, William 118–20, 127, 223
Styles, John 203
Suffolk 202–3
Sunderland 21, 35, 118
Swallow ironworks, Chapeltown
   (Sheffield) 51
Swalwell (Tyneside) 21, 186, 188, 190
Sweden 3, 6–7, 21, 24, 30–37, 43, 48, 66,
   67, 72, 77, 103, 138, 143, 146–7, 153,
   186, 191–2,
Swedenborg, Emanuel, and
   Swedenborgism 70fn, 77fn, 103
Swedish Board of Mines
   (*Bergscollegium*) 30
Swedish Ironmasters' Association 30
systems of production 7, 14–17, 20–1, 24,
   27, 32–3, 41, 43–6, 58–9, 91, 92–3,
   104–5, 108–9, 144, 176
   business management 95–8, 100–5,
      152–70, 184–92, 205
   business premises 184–91

tariffs and taxation 2, 36, 43, 52
Taylor, Joseph 223
Taylor & Maxwell 124
   Taylor, Dr Charles 124–5
textile-engineering 3, 14, 58–9, 84–5, 89,
   98–9, 103–8, 111, 141, 148, 179, 185,
   190, 198, 205–7
   See also machines and mechanization

INDEX

textile industries 8–12, 15, 20, 24, 28, 106, 143, 144, 193, 200
 cotton 25, 28, 34, 70, 87, 107–8, 124–5, 127, 135, 140, 141, 144, 181, 193, 199–201, 203–4, 207
 flax/ linens 69–70, 90, 93, 108, 118, 144, 170, 178, 182, 201
 silk 34, 146
 stocking and lace-making 12, 32
 woollen textiles 10–11, 15, 18–20, 24–5, 34, 101, 105–6, 128, 144, 170, 190, 200, 207
 worsteds 12, 17, 200
Thomas, Callisthenes 158–60, 223
Thompson, E. P. 26–7, 98, 99–100, 104
Thurgoland mills (par. Silkstone) 148, 156
tin-plate manufacture 21, 22, 24, 32fn, 227 *and see* John Hanbury
tobacco trade 19–20, 25
trading, internal 10–12, 14–15
 international 3, 18–20, 21, 24
transport infrastructure 6, 9–12, 21, 27, 35, 45, 60–1, 144, 147, 166, 176, 185–6, 207
Triewald, Mårten 138
Trowbridge (Wiltshire) 13
Tweedale, Geoffrey 50
Tyas, John, of Monmouth 159
Tyneside 12, 21, 24, 28, 33, 86, 111, 154, 176

Unitarians 88, 103, 115, 116, 120fn, 121, 166, 175, 179
 See also Cross Street chapel, Manchester; dissenting academies; Mill Hill chapel, Leeds
United Kingdom, creation of *see* Scotland: Act of Union
Universities 2, 7, 66, 83, 114, 120, 142, 175, 178–81, 204, 206
 Aberdeen 73, 180, 181
 Cambridge 73, 180
 Edinburgh 116, 121, 179, 180, 181
 Glasgow 89, 137, 169, 178, 180, 181
 Leiden 121, 179, 180
 London 121, 181
 Oxford 73, 155, 180
 Paris 121
 Philadelphia 181
Up Holland (Lancs) 20

Ure, Andrew 199

Volta, Count Alessandro 121

Wakefield 10–11, 13, 15, 20, 24, 85–7
Wakefield, Andre 65
Wales 11, 12, 35
Walker, Adam 132–3
Walker ironmasters of Grenoside and Rotherham 5, 39, 51, 52, 98, 102, 103, 104, 147, 149, 151–2, 161–4, 168–70, 176, 190, 192, 201, 204, 205, 223
 Walker, Aaron 161–2, 223
 Walker, Anne 161
 Walker, John (son of Aaron) 163
 Walker, Jonathan 161–2, 223
 Walker, Jonathan jun. 163
 Walker, Joseph (d. 1729) 161
 Walker, Joseph (son of Samuel) 163
 Walker, Joshua 163
 Walker, Samuel 155, 161–3, 223
 Walker, Samuel jun. 163
 Walker, Thomas 163–4, 168, 169
Waller, T. 109–10
Warltire, John 130, 132–3, 134, 223–4
water-power 12–13, 15, 28, 32fn, 34, 35, 38, 45–6, 51–2, 54, 58, 59, 77, 85–7, 105–7, 109–10, 112, 135–7, 147, 166, 168, 182, 185–7, 190, 200–1
 waterwheels, efficiency of types 85–6, 107, 135–6, 200–1
water-engineering 41fn, 85, 135–8, 146
waterways *see* canals and river navigation
Watt, Gregory 127
Watt, James 34, 51, 64, 70, 78, 88, 89–94, 115, 118–19, 120, 124–7, 132, 135, 137, 139, 140–1, 167, 179, 181–2, 198, 201–2, 224
Watt, James jun. 92, 124–5, 127, 139, 141, 179, 181, 224
Watts, John 159–60
Weald (area of south-east England) 35
Wearmouth Bridge, Sunderland 118, 176
Wearside 12, 86, 118, 176
Wedgwood, Josiah 78, 81, 115, 121, 133, 144, 176, 207, 224
 Wedgwood archive 78
Wednesbury (Staffordshire) 47
Wenger, Arsène 172
West Country (West of England) 11–12, 17

West Riding of Yorkshire x, 9, 22, 53, 87, 149
White, Charles 124
Whitehaven (Cumberland) 47
Whitworth, Robert 143
Wilkinson, John 45, 55, 57, 78, 91, 115–16, 119–20, 164, 224
Wilkinson, William 58fn, 224
William III, of Orange 2, 22
Wilson, Matthew 154, 156–9
Wilson, Richard G. 19
Winch, Donald 196
Winlaton (Tyneside) 21, 188
Winsor, Frederick A. 131–2
wire-drawing 32, 33, 56, 78, 150, 152, 185
Withering, William 115–16
Witton, Samuel 59
Wood, Charles 46, 47, 80–1, 180, 224–5
Wood, David 225
Wood, John 47
Wood, William 88, 114
Wootton, David 66
Wordsworth, Joshua 152, 225

workers' housing 28, 59, 144, 146, 186–8, 190–1
Wortley x, 33, 59, 78, 80, 145–52, 185, 190, 193, 204, 205
   Wortley parishes
      Ecclesfield 56, 148–50, 161–2
      Penistone 148–9
      Silkstone 148–9, 153
      Tankersley 147fn, 148–9, 152, 153
Wortley, Edward 158
Wortley Forge 54, 56, 57, 80, 154, 156–61, 168
   See also Cockshutt family; Wilson, Matthew
'wright' occupations 106
Wrigley, Joshua 225
Wyatt, John 167, 208, 220

Yarmouth, earl of 74
Yarranton, Andrew 22, 23, 176, 225
Yorkshire and Derbyshire Iron-Masters' Association 169
Yorkshire Dales 10